# MANUAL OF LOGIC CIRCUITS

PRENTICE-HALL INTERNATIONAL, INC., *London*
PRENTICE-HALL OF AUSTRALIA, PTY. LTD., *Sydney*
PRENTICE-HALL OF CANADA, LTD., *Toronto*
PRENTICE-HALL OF INDIA PRIVATE LTD., *New Delhi*
PRENTICE-HALL OF JAPAN, INC., *Tokyo*

# MANUAL OF
# LOGIC CIRCUITS

**GERALD A. MALEY**

Senior Engineer, IBM Corporation

PRENTICE-HALL, INC.
Englewood Cliffs, New Jersey

13-553503-4

Library of Congress Catalog Card Number: 74-113716

Current printing (last digit): 10  9  8  7  6  5  4  3  2  1

Printed in the United States of America

# PREFACE

This manual has been specifically prepared for the practicing logic designer or student who is interested in using or building upon previously designed switching networks. It is intended to relieve the designer of the necessity of reinventing the same functional units that have already been reworked to tedium. Hopefully, the designer will find the required functional unit—whether it be a gated latch, a ring circuit, a counter, a shift register, or an adder— contained within this manual. But failing that the designer will be given a starting point, a glimpse of what has been designed, so that his effort may commence where others have left off.

It would be unrealistic to believe that this collection is complete, or that it will ever be complete, but it is an impressive array of excellently designed functional units.

This manual contains five sections. The first two sections are concerned with Boolean Algebra, Logic Symbols, Karnaugh Mapping, and Flow Tables. In the major portion of the manual, each section is devoted to a specific logic circuit or logic connective. These sections— AND-Invert, AND-OR-Invert, and OR-Invert, Emitter Coupled Logic—represent the major areas of present logic design activity. Each of these sections opens with a presentation of design procedures that are effective when working with that connective. Combinational and sequential networks are covered in each section.

The author wishes to thank the IBM Corporation and the many IBM employees who made possible the publication of this manual.

# CONTENTS

The design of the following networks
is displayed, and their operational
characteristics are explained:

# 4 AND-OR-INVERT LOGIC, 155

PACKAGING THE AOI GATE
KARNAUGH MAPPING FOR THE AOI GATE

The design of the following networks
is displayed, and their operational
characteristics are explained:

# 5  OR-INVERT, EMITTER COUPLED LOGIC, 215

MAPPING FOR NOR
THREE STAGE NOR LOGIC
USING ECL CIRCUITS

The design of the following networks
for NOR or ECL logic is displayed,
and their operational characteristics
are explained. Following this first list
is a similar list intended for use with
NOR circuits only.

OR-INVERT, EMITTER COUPLED LOGIC, 226

OR-INVERT (NOR), 215

# APPENDICES, 285

# INDEX, 299

# MANUAL OF LOGIC CIRCUITS

# 1
# BOOLEAN ALGEBRA AND LOGIC SYMBOLS

# BOOLEAN ALGEBRA

Boolean algebra is a two valued algebra. It was originally developed for use with deductive logic where the two values corresponded to the truth and falsity of a statement. However, this two valued algebra has found many useful applications in the field of switching theory. When used as tools in switching theory, the two values are assigned to the two possible voltage levels upon which most switching networks are designed.

## Boolean Constants

Since Boolean algebra is two valued, there are only two constants; and these constants have been assigned the symbols "0" and "1." These two symbols are not to be thought of as numbers, and neither symbol is to be considered higher in rank than the other.

Boolean algebra, like ordinary algebra, allows the assignment of symbols to unknown quantities. It is customary to use capital letters from the low end of the alphabet or the high end of the alphabet for this purpose. It is understood that Boolean variables may have only two values: 0 and 1.

## Boolean Connectives

The Boolean connective "+" is read as "OR" and is defined by the following table.

| A | B | A + B |
|---|---|-------|
| 0 | 0 | 0 |
| 0 | 1 | 1 |
| 1 | 0 | 1 |
| 1 | 1 | 1 |

Expressing this table algebraically:

$$0 + 0 = 0$$
$$0 + 1 = 1$$
$$1 + 0 = 1$$
$$1 + 1 = 1$$

The Boolean connective "·" is read as AND and is defined by the following table.

| A B | A · B |
|-----|-------|
| 0 0 | 0 |
| 0 1 | 0 |
| 1 0 | 0 |
| 1 1 | 1 |

Expressing this table algebraically:

$$0 \cdot 0 = 0$$
$$0 \cdot 1 = 0$$
$$1 \cdot 0 = 0$$
$$1 \cdot 1 = 1$$

In some cases brackets are used in place of the dot, and in still other cases the lack of a symbol indicates AND.

A · B is the same as (A)(B) is the same as A B.

The Boolean connective "⁻" is read as NOT.

| A | $\overline{A}$ |
|---|---|
| 0 | 1 |
| 1 | 0 |

Many different symbols have been used for the above-mentioned connectives. The following table shows some of these symbols.

| Language term | Logic symbols |
|---------------|---------------|
| OR | $+, \vee, \cup$ |
| AND | $\cdot, \&, (\,)(\,), \cap$ |
| NOT | $^{-}, -, \sim$ |

**Theorems**

| | |
|---|---|
| I | $A \cdot 0 = 0$ |
| II | $A \cdot 1 = A$ |
| III | $A \cdot A = A$ |
| IV | $A \cdot \overline{A} = 0$ |

Each of the above theorems can be proved by setting $A = 0$ and then $A = 1$. With either value for A, the above equations will check with the definition previously given for the AND connective.

| | |
|---|---|
| V | $A + 0 = A$ |
| VI | $A + 1 = 1$ |
| VII | $A + A = A$ |
| VIII | $A + \overline{A} = 1$ |

Again, these theorems can be checked by allowing A to take on either of its two possible values, 1 or 0.

IX $$\overline{\overline{A}} = A$$

This theorem states that the inverse of $\overline{A}$ is A.

| | |
|---|---|
| X | $A \cdot B = B \cdot A$ |
| XI | $A + B = B + A$ |

Theorems X and XI state that the order of the variables with the AND connective and the OR connective is unimportant. This characteristic is called *commutativity*.

XII $$A \cdot B + A \cdot C = A(B + C)$$

Theorem XII is referred to as the factoring theorem. The theorem states that a variable common to two or more terms (a group of variables ANDed together) may be removed from each term and ANDed to the resulting expression.

XIII $$(A + B)(A + C) = A + (B \cdot C)$$

Theorem XIII is similar to theorem XII except that all AND and OR connectives have been reversed.

XIV $$A \cdot B + A \cdot \overline{B} = A$$

Theorem XIV is one of the most important theorems in Boolean algebra. Most minimization methods are based on the use of the theorem. This theorem, like all theorems in Boolean algebra, can be proved by the use of perfect induction: proving the theorem by allowing A and B to assume all possible values and all combinations of values. However, an algebraic proof is usually simpler.

*Proof:*
$$A \cdot B + A \cdot \overline{B} = A$$
$$A(B + \overline{B}) = A$$
$$A(1) = A$$
$$A = A$$

XV

*Proof:*
$$A + AB = A$$
$$A + AB = A$$
$$A(1 + B) = A$$
$$A(1) = A$$
$$A = A$$

XVI

*Proof:*
$$A + \overline{A}B = A + B$$
$$A + \overline{A}B = A + B$$
$$A(1) + \overline{A}B = A + B$$
$$A(B + \overline{B}) + \overline{A}B = A + B$$
$$AB + A\overline{B} + \overline{A}B = A + B$$
$$AB + AB + A\overline{B} + \overline{A}B = A + B$$
$$AB + A\overline{B} + AB + \overline{A}B = A + B$$
$$A(B + \overline{B}) + B(A + \overline{A}) = A + B$$
$$A(1) + B(1) = A + B$$
$$A + B = A + B$$

XVII
$$\overline{A \cdot B} = \overline{A} + \overline{B}$$

XVIII
$$\overline{A + B} = \overline{A} \cdot \overline{B}$$

These two theorems are called DeMorgan's theorems and are fundamental when working with NOR or NAND logic, as we shall see later.

*Proof by Perfect Induction of $\overline{A \cdot B} = \overline{A} + \overline{B}$*

| 1 | 2 | 3 | 4 | 5 | 6 | 7 |
|---|---|---|---|---|---|---|
| A | B | $A \cdot B$ | $\overline{A \cdot B}$ | $\overline{A}$ | $\overline{B}$ | $\overline{A} + \overline{B}$ |
| 0 | 0 | 0 | 1 | 1 | 1 | 1 |
| 0 | 1 | 0 | 1 | 1 | 0 | 1 |
| 1 | 0 | 0 | 1 | 0 | 1 | 1 |
| 1 | 1 | 1 | 0 | 0 | 0 | 0 |

This table was prepared by first listing in the left two columns all possible combinational values of A and B. Next, the 3 column was obtained by ANDing together the left two columns. The 4 column was obtained by inverting column 3, as is indicated at the top of the column, and so on across the table. Notice column 4 matches column 7 for all values of A and B and thus theorem XVIII is proved.

A summary of these and other useful theorems is given in Appendix E.

## LOGIC SYMBOLS

### Positive and Negative Logic

In the design of switching networks the symbols "1" and "0" of Boolean algebra are assigned to the two voltage or current levels employed by the electronic switching equipment to be used. When the symbol "1" is assigned to the higher of the two potentials (currents), the resultant network is said to have *positive* logic. When the symbol "0" is assigned to the higher potential (current), the network is said to have *negative* logic. The use of positive or negative logic is an arbitrary decision usually left to the designer. It is important to note that the amount of equipment required to implement a switching network is unaffected by the choice of positive or negative logic.

Consider a circuit whose output (F) is a function of two variables (A, B), and whose output and input levels are capable of assuming only +2 volts and -3 volts. Assume that the circuit behaves according to the following table of combinations.

| Inputs | | Output |
|---|---|---|
| A | B | F |
| -3 v | -3 v | -3 v |
| -3 v | +2 v | -3 v |
| +2 v | -3 v | -3 v |
| +2 v | +2 v | +2 v |

In *positive logic* the +2 v level is assigned the symbol "1" and the -3 level is assigned the symbol "0." Substitution of the logic values for the voltage levels results in the following table:

| Inputs | | Output |
|---|---|---|
| A | B | F |
| 0 | 0 | 0 |
| 0 | 1 | 0 |
| 1 | 0 | 0 |
| 1 | 1 | 1 |

This is the truth table for a two input AND function. The output of an AND is a 1 only when all inputs are 1.

In *negative logic* the –3 v level is assigned the symbol "1" and the +2 v level is assigned the symbol "0." Substitution of the logic values for the voltage levels results in the following table:

| Inputs | | Output |
|---|---|---|
| A | B | F |
| 1 | 1 | 1 |
| 1 | 0 | 1 |
| 0 | 1 | 1 |
| 0 | 0 | 0 |

This is the truth table for a two input OR function. The output of an OR is a 0 only when all inputs are 0. This statement is identical with that used for the AND connective in the previous table except 0's now replace 1's. But with negative logic, 0's and 1's are interchanged. Thus we are able to state that a positive AND and a negative OR give us the same results. This is true because a positive AND and a negative OR are two names for the same circuit.

This simple duality between positive and negative logic enables a designer to convert from one to the other very easily. To do so one redefines the symbols 1 and 0 and changes the labels on the AND circuits to OR and the labels on the OR circuits to AND. *Note*: There is no actual change in the hardware, only changes to labels are involved. Since the choice of positive or negative logic has no effect on the hardware, the author has arbitrarily selected only *positive logic* for use in this manual.

### American Standard Graphic Symbols*

Two different but overlapping graphical symbols are present in this standard: Uniform Shapes and Distinctive Shapes. For some functions there are different symbols in the two sets; for others the symbol is the same in both sets. The Uniform Shapes are shown on the left, Distinctive Shapes on the right, and symbols belonging to both sets are drawn in the center of the following pages. Symbols from the nonoverlapping parts are not permitted on the same diagram. Lines indicating input and output connections are not part of the symbol.

---

*The American Standard Graphic Symbols were obtained from the AIEE. (AIEE 91-ASA Y 32. 14–1962.) In the text material provided with the American Standard Graphic Symbols, the two possible physical conditions of each signal line are referred to as the 0-state and the 1-state. The author found this labeling very confusing when discussing state tables and state diagrams with various outputs and inputs. For this reason, the author elected to reserve the use of the word "state" for the stable conditions of a sequential circuit and substituted 0 and 1 for 0-state and 1-state.

## Symbol for AND

**Fig 1.1** AND gate

The output of an AND is a 1 if and only if all inputs are 1. An example of a three input AND (in both forms) and its associated truth is shown in Figure 1.2.

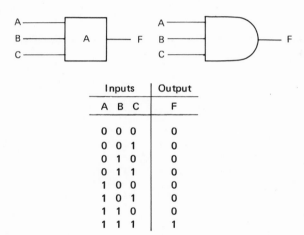

| Inputs | Output |
|:---:|:---:|
| A  B  C | F |
| 0  0  0 | 0 |
| 0  0  1 | 0 |
| 0  1  0 | 0 |
| 0  1  1 | 0 |
| 1  0  0 | 0 |
| 1  0  1 | 0 |
| 1  1  0 | 0 |
| 1  1  1 | 1 |

**Fig 1.2** AND gate with inputs and truth table

## Symbol for OR

**Fig 1.3** OR gate

The output of an OR is a 1 if one or more of the inputs is a 1.

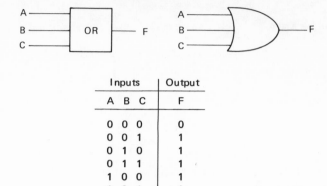

| Inputs | | | Output |
|:---:|:---:|:---:|:---:|
| A | B | C | F |
| 0 | 0 | 0 | 0 |
| 0 | 0 | 1 | 1 |
| 0 | 1 | 0 | 1 |
| 0 | 1 | 1 | 1 |
| 1 | 0 | 0 | 1 |
| 1 | 0 | 1 | 1 |
| 1 | 1 | 0 | 1 |
| 1 | 1 | 1 | 1 |

**Fig 1.4** OR gate with inputs and truth table

**Symbol for Exclusive - OR**

**Fig 1.5** Exclusive-OR gate

The output of an Exclusive-OR with two inputs is a 1 if one and only one of the inputs is a 1.

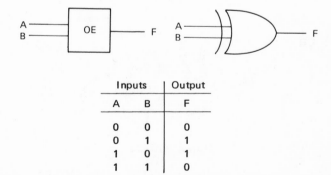

| Inputs | | Output |
|:---:|:---:|:---:|
| A | B | F |
| 0 | 0 | 0 |
| 0 | 1 | 1 |
| 1 | 0 | 1 |
| 1 | 1 | 0 |

**Fig 1.6** Exclusive-OR gate and truth table

**Symbol for Logic Negation**

**Fig 1.7**   Logic negation

The output of a logic negation is a 1 if the input is a 0, and the output is 0 if the input is 1. A small circle drawn at the point where a signal line joins a logic symbol indicates a logic negation.

An example of a logic negation applied to the output of a two input AND:

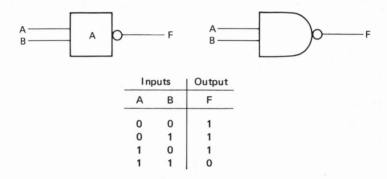

| Inputs | | Output |
|:---:|:---:|:---:|
| A | B | F |
| 0 | 0 | 1 |
| 0 | 1 | 1 |
| 1 | 0 | 1 |
| 1 | 1 | 0 |

**Fig 1.8**   Example of logic negation

**Symbol for Flip-Flop**

A *flip-flop* is a logic device that stores a single bit of information. It normally has two inputs, set (S) and clear (reset) (C), and two possible output lines, output 0 and output 1. A 1 is stored in a flip-flop when a 1 signal is applied to the set input. With a 1 stored, the 1 output will be at the 1 level and the 0 output will be at the 0 level. A 0 is stored in the flip-flop when a 1 signal is applied to the clear input. With a 0 stored, the 1 output will be at the 0 level and the 0 output will be at the 1 level. There are many versions of the flip-flop, some with additional inputs.

*Flip-Flop Complementary*

The flip-flop complementary has outputs which are always of opposite levels. This flip-flop may have an additional input, toggle (trigger) (T). The application of a 1 signal to the toggle input will reverse (change) the state of the flip-flop.

Another version of this flip-flop has in the past been called a J-K flip-flop. This flip-flop does not have a toggle input, but the state of the flip-flop is reversed (changed) if 1 signals are applied simultaneously to the set and clear inputs.

The inputs are not required to be identified with S, T, and C, as shown

**Fig 1.9**   Flip-flop

The internal numbers 1 and 0 are part of the symbol and shall be in close proximity to the 1 output; the clear input shall be in proximity to the 0 output.

*Flip-Flop Latch*

The flip-flop latch has outputs which are not necessarily of opposite states. The application of simultaneous 1 signals to the set and clear inputs causes both outputs to assume the same level for the duration of the inputs. The level of the outputs when both inputs are 1 is determined by the design of the circuit.

**Fig 1.10**   Flip-flop latch

This manual covers many more flip-flops than those just mentioned, but they are not included in the symbols approved by the American Standard Association. The Association has, however, approved the use of a rectangle for functions not elsewhere specified.

**Mixed Positive and Negative Logic**

When positive and negative logic are both used on the same diagram, it is necessary to identify as positive or negative the inputs and outputs of each logic symbol. To do this the standard calls for the use of right triangles as level indicators. The symbols shown below represent an AND with positive logic inputs (1 has been assigned to the higher potential) and a negative logic output (1 has been assigned to the lower potential).

**Fig 1.11**   AND gate with positive logic inputs

To read this diagram first examine the symbol, AND in this case. The output, therefore, is 1 if and only if both inputs are 1. The solid arrows on the input lines indicate that 1 on these lines is defined as the higher potential. The open arrowhead on the output indicates that 1 at this point is defined as the lower potential. The truth table for Figure 1.11 is shown below. "H" refers to more positive (high) potential and "L" refers to less positive (low) potential.

| Inputs | | Output |
|:---:|:---:|:---:|
| A | B | F |
| L | L | H |
| L | H | H |
| H | L | H |
| H | H | L |

(See Figure 1.12 on following page for examples of mixed positive and negative logic.)

## Symbol for Electric Inverter

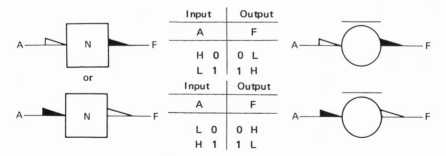

| Input | Output |
|:---:|:---:|
| A | F |
| H  0 | 0  L |
| L  1 | 1  H |

| Input | Output |
|:---:|:---:|
| A | F |
| L  0 | 0  H |
| H  1 | 1  L |

**Fig 1.13**   Electric inverter

## Single Shot

The normal (inactive) stage of the *single shot* is such that the output is 0. When activated by a significant transition of the input signal, the output will

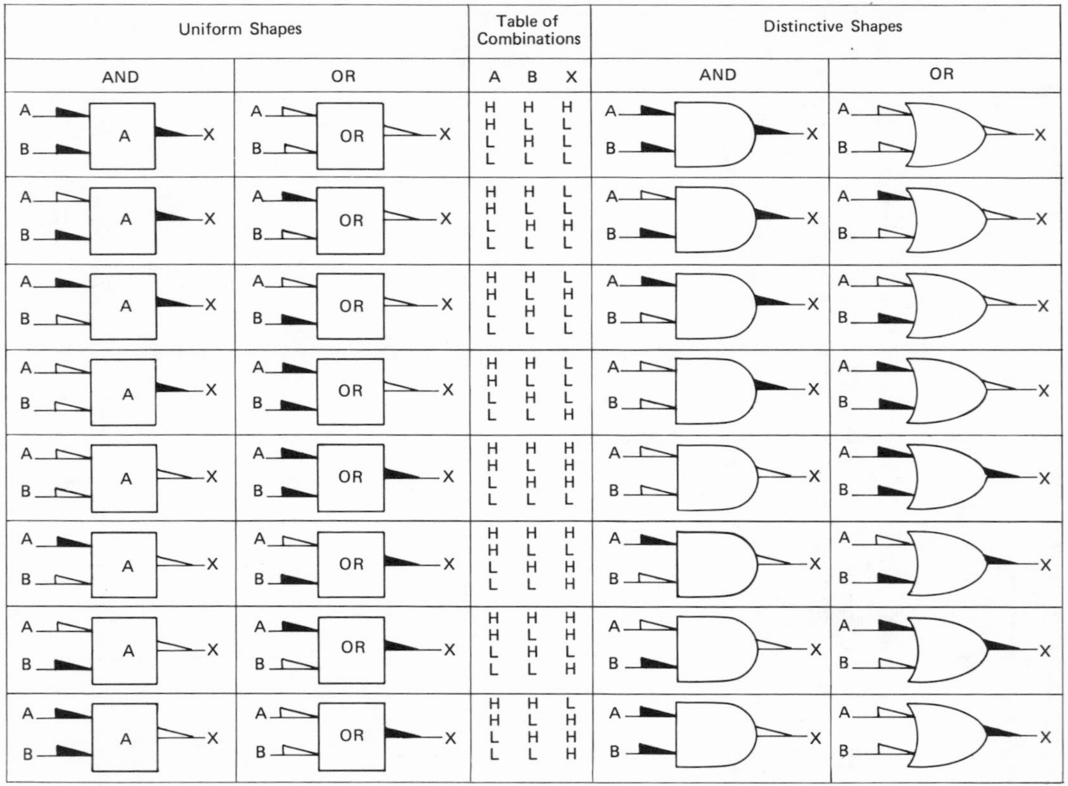

**Fig 1.12** Examples of mixed positive and negative logic

14

change to 1, remain there for the characteristic time of the device, and then return to 0 output. Output signal shape, amplitude, duration, and polarity are determined by the circuit characteristics and not by the input signal. The duration of the on time of the single shot will normally need to be included on a line inside the symbol. When required, stylized waveforms indicating duration, amplitude, and rise and fall time may be used.

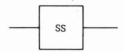

**Fig 1.14**   Single shot

## Symbol for Schmitt Trigger

A *Schmitt trigger* is activated when the input signal crosses a specified turn-on threshold towards the indicated 1 output and remains active until the input signal crosses a specified turn-off threshold towards a 0 output. Output signal shape, amplitude, and polarity are determined by the circuit characteristics and not by the input signal. The normal (inactive) stage of the Schmitt trigger provides a 0 output.

**Fig 1.15**   Schmitt trigger

## Symbol for Amplifier

**Fig 1.16**   Amplifier

This symbol represents a linear or nonlinear current or voltage amplifier. (The input and output lines are not part of the symbol.) The output of an *amplifier* is 1 when the input is 1, and 0 when the input is 0.

### Symbol for Time Delay

The duration of a time delay in a time delay unit is included with the symbol. If the delay device is tapped, the delay time with respect to the input shall be included adjacent to the tap output.

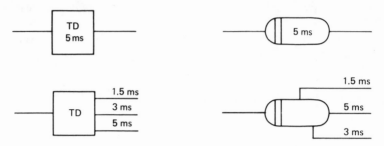

**Fig 1.17**   Time delay symbols

### Symbol for Oscillator

**Fig 1.18**   Oscillator

### Multiple Inputs

To accommodate more than three inputs, the input side of the Distinctive Shape symbol shall be extended.

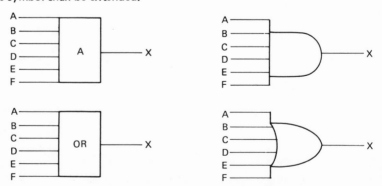

**Fig 1.19**   Examples of multiple inputs

## Use of Separate Circuits

Where circuits have the capability of being combined according to the AND (or OR) function simply by having the outputs connected, that capability shall be shown by having an additional A (or OR) inside the Uniform Shape symbol. If needed for clarity, an additional A (or OR) shall be added as a note adjacent to the connection (for the Uniform Shape). The Distinctive Shape symbol representation shall show this capability by enveloping the branched connection with a smaller sized AND or OR symbol.

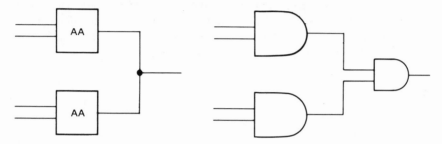

**Fig 1.20**   Separate input usage

## Extension of Inputs

Where a circuit is used to add inputs to another AND or OR circuit and the connection from this second circuit to the first is made at other than a normal input or output of the first circuit, the connection will be shown without polarity and will be labeled E (for *extender*). The relative position of the symbols and of the interconnecting extender line in the example below is not intended to be significant, and it may be drawn as any other input.

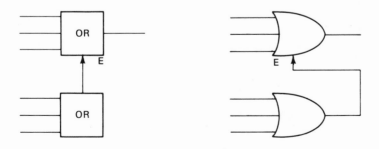

**Fig 1.21**   Input extensions

**Presentation Techniques**

1. Symbol orientation does not affect the meaning of the symbol.
2. Symbol line thickness does not affect the meaning of the symbol.
3. Symbol size does not affect the meaning of the symbol.
4. Signal flow direction—While logic diagrams may completely indicate direction of signal flow by the symbols themselves, arrowheads superimposed on lines, an extended border on the input side of a rectangle, or a convention of unidirectional flow suitably noted on the diagram may be used where required for clarity. No input shall enter a symbol in the vicinity of an output or on the output side of a symbol. Inputs may be drawn on any side of the symbol (except the output side), although the side opposite the output is preferred.

# 2
# KARNAUGH
# MAPS
# AND
# FLOW
# TABLES

# KARNAUGH MAPPING

## Introduction

This manual contains a great many complex logic networks that may be used without modifications or alterations. But the value of the manual can be greatly increased if the user is able to modify or tailor the listed networks to his specific needs. This ability to modify requires a gate by gate understanding of the operating networks. The problem is how to convey this information clearly and concisely. The author has elected to use Karnaugh maps and flow tables for this purpose. However, this requires the user to have a working knowledge of maps and flow charts which are not at all standardized in the industry. This fact becomes immediately clear when one tries to explain a logic network where the input signal passes through more than two gates in series.

To insure complete communication between author and reader, no previous knowledge of mapping or flow charting is assumed. The author covers first the fundamentals of Karnaugh mapping and flow charting. Later these fundamentals are used to develop specific techniques for handling NAND, NOR, and AND-OR-Invert gates. The major sections of this manual are restricted to particular logic connectives, and the specific techniques for designing with these connectives are found at the front of each section.

## Truth Tables

The simplest method of explaining the terminal operation of a combinational network (combinational meaning one without memory) is by way of a truth table. The truth table lists all possible input combinations, and next to each is the output signal that is being generated. (For example, see Figure 2.1.)

| Inputs | Output |
|--------|--------|
| A B C  | F      |
| 0 0 0  | 0      |
| 0 0 1  | 1      |
| 0 1 0  | 1      |
| 0 1 1  | 0      |
| 1 0 0  | 1      |
| 1 0 1  | 0      |
| 1 1 0  | 0      |
| 1 1 1  | 1      |

**Fig 2.1**  Three variable truth table

21

Truth tables can be used for problems with any number of input variables, and their value rests in their uniqueness. Unlike Boolean expressions, there exists one and only one truth table for any specified function. However, truth tables give no information concerning the internal gating of the network.

### The Two Variable Map

A two variable Karnaugh map has four cells, one for each of the four rows of a two variable truth table. Each row corresponds to a specific cell, as shown in Figures 2.2 and 2.3.

To transfer a function from a truth table to a map, each entry in the output column of the table is placed in the appropriate cell of the map. The function

| Inputs | | Output |
|---|---|---|
| A | B | F |
| 0 | 0 | 0 |
| 0 | 1 | 1 |
| 1 | 0 | 0 |
| 1 | 1 | 0 |

**Fig 2.2** Two variable truth table

**Fig 2.3** Karnaugh map

shown has only a single output of 1, and this 1 is entered into the upper right-hand cell of the map. This cell has input conditions of A = 0, B = 1; and that is the row in the truth table that specifies an output of 1.

### The Three Variable Map

A three variable Karnaugh map has eight cells, one for each of the eight rows of the three variable truth table. The eight cells are arranged in four rows of two cells each. The important point to note is the labeling of the rows.

The rows of the map are labeled so that as one moves up or down within a column, only one input variable need change at a time to identify the row. As a

| Inputs | Output |
|--------|--------|
| A  B  C | F |
| 0  0  0 | 0 |
| 0  0  1 | 1 |
| 0  1  0 | 1 |
| 0  1  1 | 0 |
| 1  0  0 | 1 |
| 1  0  1 | 0 |
| 1  1  0 | 1 |
| 1  1  1 | 0 |

**Fig 2.4**   Three variable truth table

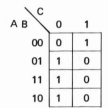

**Fig 2.5**   Karnaugh map

matter of fact, the entire array of cells in any Karnaugh map is always arranged so that as one moves vertically or horizontally one cell at a time, only one identifying input variable will change. Cell A = 1, B = 1, C = 0 is next to cell A = 0, B = 1, C = 0. The top row of cells is also considered to be adjacent to the bottom row of cells. The three variable map would have to be drawn as a cylinder to show this adjacency. Figures 2.4 and 2.5 show a three variable truth table and Karnaugh mapping. It is understood that cell A = 0, B = 0, C = 0 is next to cell A = 1, B = 0, C = 0; and cell A = 0, B = 0, C = 1 is next to cell A = 1, B = 0, C = 1.

**The Four Variable Map**

Since there are 16 rows in the truth table of a four variable expression there are also 16 cells in a four variable map. It should be noted that each time the number of input variables is increased by one, the number of required cells doubles. It is this doubling that makes it impractical to use Karnaugh maps for large numbers of variables (see Figure 2.6).

Once again, observe both horizontal and vertical labeling. As one moves from any cell to any other adjacent cell, diagonal movement is not included—only one variable changes. The right-hand column is considered to be adjacent to the left-hand column just as the top and bottom rows are considered to be adjacent.

|        | C D |    |    |    |
| A B    | 00  | 01 | 11 | 10 |
| :----: | :-: | :-: | :-: | :-: |
| 00     | 0   | 0  | 0  | 0  |
| 01     | 0   | 1  | 0  | 0  |
| 11     | 0   | 0  | 0  | 1  |
| 10     | 0   | 0  | 0  | 0  |

**Fig 2.6**  Four variable Karnaugh map with a 1 in cell
A = 0, B = 1, C = 0, D = 1 and cell A = 1, B = 1, C = 1, D = 0

### Use of Adjacencies

Most Boolean minimization methods are based on the repeated application of one basic Boolean theorem. This theorem states:

If two terms are identical except for only one variable which is in true form in one term but inverted in the other, then both terms can be replaced by a common factor. Expressing this algebraically:

$$\text{Example 1} \quad A\overline{B} + \overline{A}\overline{B} = \overline{B}$$
$$\text{Example 2} \quad AB\overline{C}D + ABCD = ABD$$

The Karnaugh map has been specifically designed to provide a visual application of this basic theorem. Cells have been labeled so that the theorem applies between any two adjacent cells, whether they are horizontally adjacent or vertically adjacent. Thus when two adjacent 1's are found in a Karnaugh map,

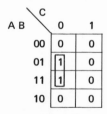

**Fig 2.7**  Map with two adjacent 1's

there must be a single term that will cover both. A loop is usually drawn around the adjacent 1's to show that they are read from the map as a common term (see Figure 2.7). In this example the loop is read from the map as $B\overline{C}$. The variable A which appeared in true form in one term and in inverted form in the other single term is dropped from the new common term, leaving only $B\overline{C}$.

**Four Cell Loops**

It has been demonstrated that two adjacent 1's can be read from the map as a single term, and likewise a group of four 1's can also be read as a single term. A group of four 1's must, however, form a 2 x 2 square of cells or a straight line of four cells, as shown in Figures 2.8 and 2.9.

Figure 2.8 has a four cell loop which is read as B, while Figure 2.9 is read as

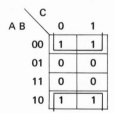

|     | C  |     |
|-----|-----|-----|
| A B | 0 | 1 |
| 00 | 0 | 0 |
| 01 | 1 | 1 |
| 11 | 1 | 1 |
| 10 | 0 | 0 |

**Fig 2.8**   Map with four cell loop

|     | C  |     |
|-----|-----|-----|
| A B | 0 | 1 |
| 00 | 0 | 1 |
| 01 | 0 | 1 |
| 11 | 0 | 1 |
| 10 | 0 | 1 |

**Fig 2.9**   Map with four cell loop

|     | C  |     |
|-----|-----|-----|
| A B | 0 | 1 |
| 00 | 1 | 1 |
| 01 | 0 | 0 |
| 11 | 0 | 0 |
| 10 | 1 | 1 |

**Fig 2.10**   Map with four cell loop

C. If the top two cells and the bottom two cells of a three variable map contain 1's, as shown in Figure 2. 10, then this loop is read as $\overline{B}$.

**Loops on Four Variable Maps**

Loops of adjacent 1's on a four variable map may contain one, two, four, or eight cells. The loops of eight are found in 2 x 4 patterns. Loops of two, four,

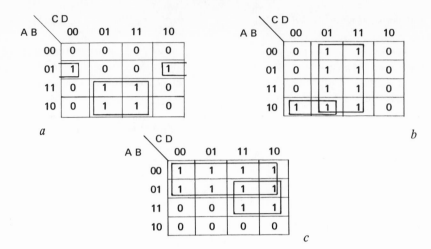

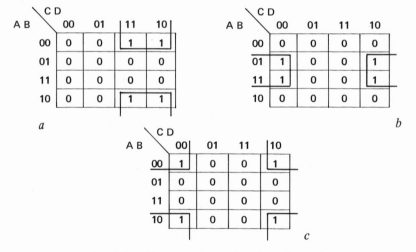

Fig 2.11   Four variable maps with loops

Fig 2.12   Karnaugh maps showing adjacencies

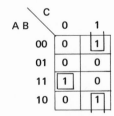

Fig 2.13   Map with two loops

and eight cells are shown in Figures 2.11(a), (b), and (c), and 2.12(a), (b), and (c).

In Figure 2.11 the loops on map (a) are read as $\overline{A}B\overline{D}$ and AD. The function for the entire map is read as $F = \overline{A}B\overline{D} + AD$. Map (b) is read: $F = D + A\overline{B}\overline{C}$. Map (c) is read as $F = \overline{A} + BC$. Notice that the loops are always made as large as possible, even though they overlap other loops. Because the top and bottom rows of the map are adjacent and the left and right columns are adjacent, some loops are not so easily recognized.

In Figure 2.12 the loop on map (a) is $\overline{B}C$; on map (b), $B\overline{D}$; and on map (c), $\overline{B}\overline{D}$. For example, see Figure 2.13. In this case we have two loops:

$$\text{loop of 1} \quad AB\overline{C}$$
$$\text{vertical loop (split)} \quad \overline{B}C$$

A loop was drawn around the lone 1 to indicate that it had no adjacent 1's and must be read from the map as a term by itself.

## Using the Karnaugh Map

The Karnaugh map is most often used to obtain a *sum of products* logic network: a logic network comprising a minimum column of AND gates connected to a common OR gate. Such a network is obtained from a direct

| Inputs |  | | Output |
|---|---|---|---|
| A | B | C | F |
| 0 | 0 | 0 | 0 |
| 0 | 0 | 1 | 1 |
| 0 | 1 | 0 | 0 |
| 0 | 1 | 1 | 1 |
| 1 | 0 | 0 | 0 |
| 1 | 0 | 1 | 0 |
| 1 | 1 | 0 | 1 |
| 1 | 1 | 1 | 1 |

**Fig 2.14** Truth table, map, and logic network

reading of the map. Each loop will correspond to an AND gate and the inputs to these ANDs are obtained by reading the loop identification from the map (see Figure 2.14).

Since the implementation requires an AND gate for each loop, a minimum number of loops should be used. Every 1 in the map must be enclosed in a loop, but it is not necessary to enclose a 1 more than once. In Figure 2.14 a third loop could have been drawn around cell 011 and 111, giving use to the term BC. However, this loop (AND gate) is redundant, since these two cells have been covered by the other loops.

The size of the loops is important, for small loops waste inputs to AND gates—see Figure 2.15.

Notice that the two loops have been made as large as possible, even to the point of overlapping each other. Either of these loops of four could have been reduced to a loop of two and yet maintained complete coverage of all 1's. But a loop of two requires one more input to an AND gate than a loop of four; BD could have been $\overline{A}$BD to no advantage. For this reason loops should always be made as large as possible.

The drawing of the loops should be performed in an organized manner to avoid redundancies.

*First*: Look for single 1's that cannot be enclosed in a loop of two. Draw a loop around each
*Second*: Look for groups of two adjacent 1's that cannot be enclosed in a loop of four. Draw a loop around each group of two.
*Third*: Continue this search for even larger loops until the map area has been exhausted.

Choosing a set of loops should also be performed in an organized manner.

*First*: Look for 1's that are contained in only one loop. Record each of these loops by their identifying variables. These terms are called *prime implicants*.
*Second*: Choose a minimum set of loops to cover the remaining 1's. Record these terms, which are called *implicants*. There may be more than one set of minimum implicants which means there may be more than one minimum sum of products solution (see Figure 2.16).

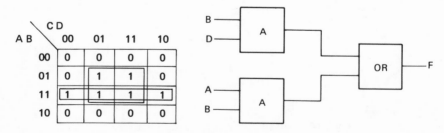

**Fig 2.15** Map and logic network

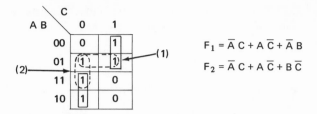

$$F_1 = \overline{A}\,C + A\,\overline{C} + \overline{A}\,B$$

$$F_2 = \overline{A}\,C + A\,\overline{C} + B\,\overline{C}$$

**Fig 2.16**    Map with two equal solutions

## FLOW TABLES

This manual contains more than one hundred complex sequential networks. These networks, as the name sequential implies, have memory capability. Their output signals are functions not only of present input signals but of past inputs as well. A counter is a good example of a sequential network. In the past, explanation of these networks has hinged on the use of timing charts. These charts depicted the rise and fall of the input signals and displayed the resulting output patterns. Charts of this type are useful, and many of the enclosed networks are supplied with them; but such charts usually lack generality. They are drawn to explain the operation of the network under one pattern of input signals but are impractical when it comes to network operations under all possible input sequences. For explanations of this type, word statements or a *flow table* may be used. To avoid the ambiguity often present in word statements, the author has elected to use flow tables. Unfortunately, the industry has not arrived at a standardized table. The type of table used in this manual is basically that which was first proposed in a doctoral dissertation by D. A. Huffman in 1953. Before proceeding with the use of these tables, it is important to understand the concept of stability and nonstability as it applies to a sequential network.

### Stability and Nonstability

The key to the analysis and design of sequential networks is the understanding of stability or lack of stability associated with each logic gate. Any gate, OR, AND, NAND, etc., is said to be stable if the output value of that gate corresponds to what would be expected from an examination of the inputs to that gate and its associated truth table. When the network agrees with its truth table it is stable. This is hardly surprising, since this is the mode of operation most often encountered. But logic gates are far from ideal and all delay a propagating signal. The delay may be very small but it is always present. This delay introduces inconsistencies to our previously explained truth tables. There may now be short periods of time when the truth table for a gate is violated. These violations can occur when an input or a set of inputs change and this change has not had time to affect the output. This period of truth table violation is called a period of nonstability. The gate cannot remain for long in a nonstable condition, for the inherent delay of the gate is finite and the output will soon match the value decreed in the truth table.

### Network Stability and Nonstability

A network is said to be stable when all gates contained in that network are stable. And likewise when any gate in a logic network is nonstable, the entire network is said to be nonstable. But unlike a logic gate, which when given enough time will become stable, a logic network may move from one state of nonstability to another state of nonstability. An oscillator is an example of a network that never does reach a stable state but continually moves from one nonstable state to another nonstable state and back again.

### The Construction of a Flow Chart

Any gate within a sequential network can have only two possible types of inputs: from other gates and inputs from external sources. Any gate can have either type of input or both types. The prime function of a flow chart is to show all possible input values to all gates contained in a network. A flow chart, therefore, contains one square (like a Karnaugh map) for each possible combination of values on all gate inputs, whether they be internally or externally generated. A network having two external inputs and two gates will,

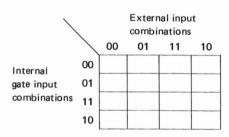

**Fig 2.17** Flow table

therefore, have a primitive flow chart containing 16 squares. There are four lines, two external and two internal, and each can be at a 0 level or a 1 level, giving $2^4$ combinations of input values. The flow chart is drawn so that all external combinations are displayed across the top of the table and internal combinations are shown down the side of the chart. See Figure 2.17 and notice that this figure is identical to a four variable map, and except for the specific input labeling, this relationship to the Karnaugh map will always exist.

## Network Analysis by Flow Chart

The use of the flow chart for network analysis will be explained by way of an example. A flip-flop constructed from two NOR gates will be used (see Figure 2.18).

1. Label all gates with decimal numbers.
2. Write Boolean expressions for the output of each gate.

$$\text{Output gate } 1 = \overline{R + 2} = \overline{R}\ \overline{2}$$
$$\text{Output gate } 2 = \overline{S + 1} = \overline{S}\overline{1}$$

   Notice that it was not necessary to look any further than to the immediate inputs to a gate to be able to write the gate output expression.
3. Draw the flow chart, including one column for each external input combination and one row for each internal input combination. In this

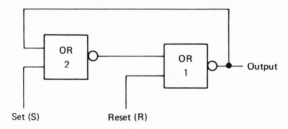

**Fig 2.18**   NOR flip-flop

|  1 2 | S R | | | |
|---|---|---|---|---|
| | 00 | 01 | 11 | 10 |
| 00 | 1 | 0 | 0 | 1 |
| 01 | 0 | 0 | 0 | 0 |
| 11 | 0 | 0 | 0 | 0 |
| 10 | 1 | 0 | 0 | 1 |

**Fig 2.19**   Incomplete flow chart for flip-flop

example there are two external inputs—S and R; and therefore, there will be four columns. There are two gates and thus four rows (Figure 2.19).
4. Using the chart as a map, enter the Boolean expression for the output of gate 1. Record the 1's and 0's in the left section of each square (left because the labeling of the chart has the gate 1 on the left). See Figure 2.19.

| 1 2 | S R 00 | 01 | 11 | 10 |
|-----|------|----|----|----|
| 00 | 11 | 01 | 00 | 10 |
| 01 | 01 | 01 | 00 | 00 |
| 11 | 00 | 00 | 00 | 00 |
| 10 | 10 | 00 | 00 | 10 |

**Fig 2.20**  Developing flow chart for flip-flop

5. Now enter on map Boolean expression for output of gate 2. Mark 1's and 0's to right of previous 1's and 0's. See Figure 2.20.
6. At this point stable and nonstable states can be identified. When a binary number within a square is equal to the binary number labeling that row, then the network is stable. This definition of stability is precisely that which was explained earlier. When the inputs to all gates (row labels) match the values of the truth tables for the output of all gates (numbers in squares), then the network is stable. The stable states are now circled and given identifying numbers, as in Figures 2.21(a) and 2.21(b).

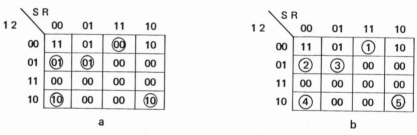

**Fig 2.21**  Flow chart with stable states

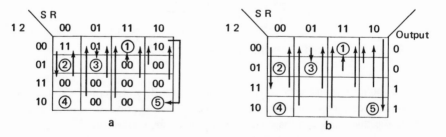

**Fig 2.22**  Flow chart for flip-flop

7. All squares that are not labeled as stable states are nonstable states. As explained, a network cannot stay for long in a nonstable state; and the binary number within the square indicates how the network will move, assuming there is no external input change at this time. The binary number within the square indicates the row to which the operating point will move. The movement is always within the same column, since external inputs are assumed not to change at this time. The operating point movement is indicated on the flow chart by the use of arrows—see Figures 2.22(a) and 2.22(b). The binary numbers are removed from the chart—as in Figure 2.22(b)—to clarify the drawing. The output signal which must be generated from an internal gate—gate 1 in this example—is also relisted to the right of the chart for convenience. This relisting involves only the moving of a column of 1's and 0's from the left of the map to the right of the map. If the network generates two outputs, then the appropriate columns are moved again; this is also for convenience only.

Flow charts can be drawn for any sequential network. The method just demonstrated gives a complete chart for operation of all gates within the network. Charts of this type are usually large since the number of rows is equal to $2^n$ when n is the number of gates. For example, a five gate problem has 32 rows. These basic or primitive charts may be needed if one is concerned with individual gate operation, as the designer always is. But the user may not require such detail, and for this reason more compact flow charts will be developed later in this section.

## The Operating Point

The *operating point* of a network is actually the state of the network at any given instant of time. It may be thought of as a pointer on a given flow chart that points at a set of conditions. The pointer or operating point moves from column to column as the input conditions change, and it moves up and down within the columns according to arrows drawn in the chart. The operating point continues to follow the arrows in a given column until a stable state is reached. Having arrived at a stable state, the operating point will remain there until an input condition moves it to another column. Here it will follow the arrows again until a stable state in that column is reached.

Let us return to the completed flow chart for the NOR flip-flop in Figure 2.22(b). Notice that with both inputs (S R) down, the network has two stable states (② and ④). Stable state ② has an output of 0, while state ④ has an output of 1. Let us assume that both inputs are down and the network is at state ② .

Now assume that the set line is raised and, after a given time interval, lowered again. As the set line is raised the operating point will move from ② in column 00 horizontally to column 10. In this column the operating point will now follow the arrows to the top row and then to ⑤. It will remain at stable state ⑤ until the set line is lowered, at which time it will move horizontally to ④ in column 00. Since the network is stable, there will be no further movement until another input change occurs. Notice how the output signal changed throughout this movement of the operating point. It started at 0, changed to 1 at ⑤, and remained 1 even after the set line was lowered. Raising the reset line will cause the output to return to 0 by moving the operating point to column 01 and stable state ③ . When the reset line is lowered to 0, the operating point will move to ② where the output is also 0.

Flow charts can be used to answer all types of questions concerning operation of the network. As an example, the answers to the following questions concerning the NOR flip-flop can be obtained from the flow chart.

*Question 1:* What is the output of the flip-flop if both set and reset lines are up at the same time?
*Answer:* Under an input condition of 11 the network moves to a ① where the output is a 0.
*Question 2:* What happens to the network if both inputs are now lowered at the same time?

*Answer:* Under this change of inputs the operating point will move from ① to the upper left-hand square, column 00. From here it will try to follow the arrows to row 11 and then back to row 00. It is important to note that the network will fail in this oscillator motion, since gate 1 and gate 2 are required to change at exactly the same time to go from row 00 to row 11. Two gates will never change at the same speed. What will happen in this case is that the operating point will start to move from row 00 as gates 1 and 2 start to rise, but one gate will arrive at a 1 level ahead of the other. In this way the network will move to either row 01 or 10; and once it reaches either of these rows it will lock there, never reaching row 11. This type of operation is called a *critical race condition* and dependence upon the outcome of such races should be avoided.

## Races—Critical and Noncritical

In our previous example a critical race condition was encountered. As was stated, the dependence upon the outcome of critical races should be avoided. Notice that avoidance of the network is not advised as many very useful networks contain races of this type. Rather, avoid those changes of input conditions that lead to critical races. The author has removed all arrows from squares in flow charts that have critical race conditions associated with them and has drawn a dash in these squares. While not advised by the author, there are many tricks that can be used to aid in the outcome of these races. The loading of one gate, adding extra wiring delay, or modifying a transistor characteristic can be used reliably in many cases to determine the outcome of a critical race. The necessity for tricks of this type can usually be avoided by good design.

*Noncritical races* are race conditions within a network where the final outcome of the race is predictable. Notice how one gate changes at a time as the operation point moves from one adjacent row to the next. Thus, any arrow that points to an adjacent row (top and bottom rows are adjacent) gives no problem as far as race conditions are concerned. But arrows that skip rows produce race conditions, since more than one gate must change at the same time to cause the operating point to follow the arrow exactly. Such race conditions may be

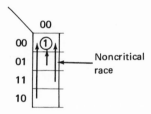

**Fig 2.23**   Partial flow chart with noncritical race

critical, as in the past example, or noncritical. A noncritical race is one in which the outcome is always the same regardless of which path the operating point takes. The usual example of a noncritical race is that in Figure 2.23 which shows only one column of a flow chart. The race condition from row 11 to 00 is noncritical, since the network may move to row 01 or 10 as it tries for row 11. The reason this network may move to row 01 or 10 is that it will only be redirected from either of these rows back to row 11. The critical condition occurs when either of the intermediate points misdirects or locks the network in an undesired state.

**Compressed Flow Charts**

While the above-mentioned flow charts are extremely detailed, they are also very large and complicated. They supply more information than is needed in most cases, assuming that the networks under study have been carefully designed and previously checked. A compressed or merged flow chart which shows only the operation of selected lines within the network is sufficient for most applications. Charts of this type are used throughout the manual and are obtained by writing Boolean expressions for selected internal lines rather than for each line. In the previous example two Boolean expressions were used as the starting point for the analysis of a NOR flip-flop. A much simpler analysis of the same network can be obtained by writing only one Boolean expression, the expression for the output line (gate 1)—see Figure 2.18.

$$\begin{aligned}
\text{Output gate 1} &= \overline{R + \overline{(S + 1)}} \\
&= \overline{R + \overline{S}\overline{1}} \\
&= \overline{R}(S + 1) \\
&= \overline{R}S + \overline{R}1
\end{aligned}$$

Notice how this one expression encompasses all gates. It is a more complex expression than either of the two replaced expressions, and in some ways it is less precise. A compound expression of this type makes the assumption that all gate delays can be lumped together at one point, the output of gate 1. Of course this is not true, but it is accurate enough for simple network analysis. This new Boolean expression is now charted in the usual manner. The chart will have only two rows in this case, since only one expression is required (see Figure 2.24).

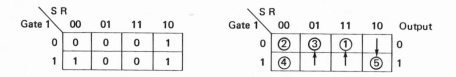

**Fig 2.24**  Compressed flow chart for NOR flip-flop

This compressed flow chart should be compared with that of Figure 2.22. Notice that it has the same number of stable states with the same related output levels. But also observe how the race condition in column 00 no longer appears. It is this masking of race condition that makes it necessary to completely chart every untried network. In this particular case, the masking of a race condition is of little importance, since the only path to this race is from column 11 (state ①) to column 00. This in itself requires a double change of input conditions. As stated earlier, one can never count on two signals changing at the same time; and therefore the user of this network should not be planning on going from ① to②. With this restriction on change of inputs in mind, the compressed flow chart is adequate for analysis of the network.

The author of this manual has constructed a primitive flow chart for each of the networks in this manual to determine if race conditions exist. From these exhaustive charts the author has selected various lines within the network that are essential to the understanding of the network. These are the lines that have been flow charted and displayed. The author has been careful not to mask, by oversimplifying, internal gate conditions that would be of importance to the user.

# 3

# AND-INVERT
# LOGIC

# AND-INVERT

## Combinational Logic—General Information

The first portion of the AND-Invert (NAND) section of this manual deals with combinational logic only. That is, logic networks which have no feedback paths and thus no memory. The outputs of these networks are functions of present input conditions only. Later in this same section many useful sequential networks are discussed.

Combinational networks are often compared for speed and component count against a sum of products solution. What is meant here is the NAND version of the sum of products solution. This NAND sum of products solution can be obtained by the following procedure:

1. Karnaugh map the desired function.
2. Select a minimal set of implicants.
3. Draw AND-OR network indicated by the selected set of implicants. (This is the normal sum of products network.)
4. Relabel *all* logic gates to NAND.
5. Invert all single variable inputs that feed the output NAND gate.

*Example*

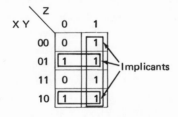

Output = $Z + \overline{X}\,Y + X\,\overline{Y}$

*Sum of Products Network*

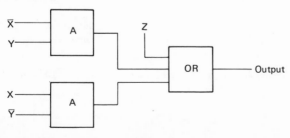

## NAND Sum of Products Network

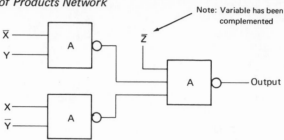

Many of the following circuits are referred to as minimum solutions. The classic definition of minimality is being used here. First, there is no network that uses fewer gates, although there may be many networks that have the same number of gates. Second, of the solutions with equal numbers of gates, the solution with the lowest number of total inputs to gates has been selected. The reader should take care in noting the conditions under which this minimality is claimed. The availability or lack of availability of input complements, as an example, is extremely important when referring to minimality. The author wishes to point out that he has not discovered a method of proving minimality in logic networks containing four, five, six stages of gating. Minimality in these cases has been obtained by what the author believes to be exhaustive trial and error computer runs.

As we have seen, NAND sum of products networks can be obtained from a direct reading of a Karnaugh map. While networks of this type are interesting, their usefulness is based on one rather unrealistic assumption: all input variables are available in true and complemented form. This is seldom the case. If NAND sum of products solutions are used, the designer will find that he requires many inverters. However, this inverter problem can be avoided to a large extent by designing three stage networks. A three stage network is one in which input signals may pass through three gates in series rather than the two gates of a sum of products solution. The procedure for developing a three stage network uses the Karnaugh map as a basic tool. The procedure is similar to that required for a sum of products solution; but some modifications have been made, as is shown below in an illustrated example problem.

| *Step* | *Example Problem* |
|---|---|
| | Output = $X \bar{Y} + \bar{X} Y$; input complement not available |
| 1. Draw Karnaugh map of desired function. | |

|   | Y |   |
|---|---|---|
| X | 0 | 1 |
| 0 | 0 | 1 |
| 1 | 1 | 0 |

2. Indicate on map the square or squares that represent the available input variables.

Only X and Y are available

3. *All* loops drawn on the map *must* include a marked square. Draw inplicant loops on the map to cover all 1's. Be sure that every loop covers a marked square, even if it must enclose some 0's.

4. The loops are temporarily treated as a normal set of implicants and a network is constructed. One NAND block is required for each loop plus one additional NAND for the output. The input lines to the NANDs are obtained by referring to the loops on the map.

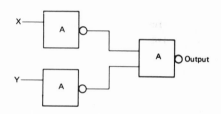

5. At this point the network is correct except for the 0's that have been enclosed in the loops. These 0's must be inhibited. This is accomplished by considering these 0's as a new problem and mapping them as 1's on a new map.

6. Again, draw implicant loops on the map to cover all 1's. Be sure every loop covers a marked square. It should not be necessary to enclose any 0's at this point.

7. A NAND block is drawn for each of these loops. The inputs to these NANDs are taken from the loops on the map.

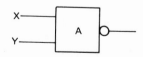

8. The output of these blocks may be
   thought of as inhibit terms. Each
   NAND is related to a specific 0 or set
   of 0's that was previously enclosed in
   a loop. The outputs of these blocks
   are selectively connected to the
   previously designed NANDs. Be sure
   each 0 in each loop is inhibited. If a
   0 is enclosed by more than one loop,
   it must be inhibited from each of the
   enclosing loops.

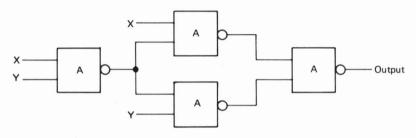

The above procedure, like sum of products Karnaugh mapping, may produce
several alternate but equal solutions. This situation is especially noticeable when
some input variables are available in both true and complemented form.

While three stage solutions are important, they are no more important than
those solutions that require four, five, or six stages of gates. Unfortunately, there
is no deterministic procedure for obtaining these solutions. The concept of
blocking a 0 from a loop is often advantageous, but which 0 to block on which
stage is but a guess at this point.

In explaining the operations of multistage networks the author uses this
blocking or inhibiting concept, but it would be misleading for the author to
imply that these networks were obtained from a simple procedure. In many
cases these Karnaugh maps were drawn after the network was designed by trial
and error.

## Karnaugh Maps and Subfunctions

As explained, Karnaugh maps are drawn above each of the combinational
networks in the following sections. The loops nearest the output lines on the
maps show the prime implicants that are being used to generate the desired
output functions. Each loop on this map will relate to a NAND gate directly
below it. Some of these prime implicants may enclose 0's that must be blocked
in the next stage of logic. Thus these enclosed 0's are remapped as 1's in a map

to the left of the first. The 1's in this map are now looped, and once again each loop is related to a NAND gate directly below. If 0's are enclosed, the procedure is started once again to the left of the last map, and so on.

In addition to aiding in understanding the operation of a network, the Karnaugh maps drawn above each network can be useful when searching for particular subfunctions. This is brought about by the fact that each NAND gate relates directly to a loop on a map, and more specifically, it relates only to the 1's in that loop. Thus, the output of each NAND generates a function which when *complemented* will contain all the terms indicated by 1's within the related loop. When a loop is found that contains a desired subfunction or portion of a subfunction, then the output of the related NAND is connected as an implicant to the new network.

*Example:* From the Exclusive-OR network previously designed, generate a subfunction that has $\overline{X}\,Y$ as one of its implicants.

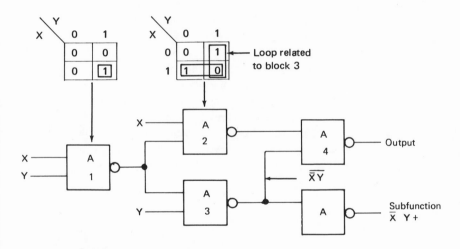

Karnaugh maps are also used in this manual to replace Boolean expressions wherever possible. The reason for this rather unusual method of expression is due to the uniqueness of the mapped function. There are many ways to write a Boolean function, but only one way to map it. Thus, maps that contain no loops are used to show the output function only and do not reflect on the design procedure or availability of subfunctions.

## The Wired Function

The networks shown in this section of the manual make no use of the *wired* or *dotted* function. "Wired" and "dotted" are words used to explain the ability to produce an additional logic connective by shorting together the outputs of two or more logic connectives. Some circuits will produce a "wired OR" when two or more outputs are shorted, while others produce a "wired AND." No circuit has the ability to perform both types of dotting. The ability to dot and the type of connective formed is a function of the circuit being used.

The wired OR is best thought of as a method of extending the total number of inputs connected to a block. If two blocks with three inputs each have their outputs shorted together, the output function is the same as that which would be obtained from a six input NAND.

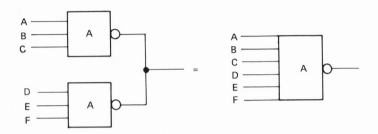

While this ability to extend the total number of inputs is important, it does not show up in this section of the manual, since no input limitations were placed on the logic blocks being used. The author assumes the designer will extend the number of inputs by this method when and if it is appropriate.

The wired AND is best thought of as a technique to save the number of inputs required of a block. The wired AND allows the shorting together of two or more outputs that are connected to a single block, as shown below:

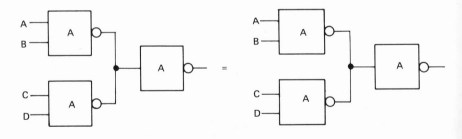

This saving of the number of inputs must be performed with care, for it does away with the ability to use the output of either dotted block separately. The networks in this section of the manual do not use this technique, for it is not always available and it is a technique that is applied after the network is designed. The designed network is scanned and wired ANDs are incorporated where subfunctions are not being used. The author assumes the designer will incorporate this capability where appropriate.

## NAND COMBINATIONAL NETWORKS

### Exclusive-OR No. 1

The following network is the sum of products solution for the two input Exclusive-OR problem. This network has a minimal signal delay of three logic blocks, assuming that input complements are not available. If input complements are available, then the two single input blocks may be removed, reducing the signal delay to two logic blocks and the block count to three.

*Karnaugh Map*

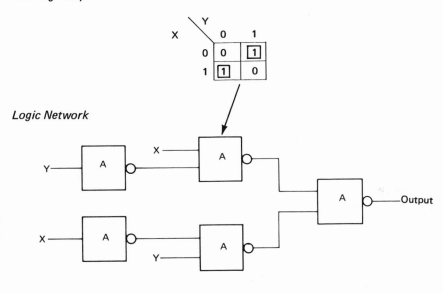

*Logic Network*

Output = $X\overline{Y} + \overline{X}Y$

*Operation*

The output will be up if one and only one of the inputs is up.

## Exclusive-OR No. 2

When input complements are not available, the following Exclusive-OR network is highly recommended. It requires only three logic blocks and has minimal signal delay.

*Karnaugh Mappings*

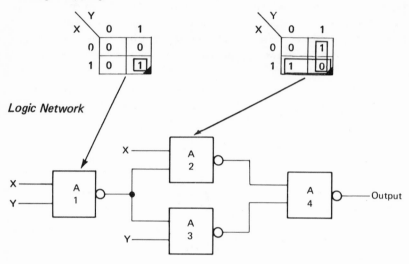

*Logic Network*

Output = $X \overline{Y} + \overline{X} Y$

*Operation*

Block 1 is used to inhibit the term XY from blocks 2 and 3.

**Majority Circuit (Two or Three out of Three Circuit)**

The sum of products solution to the majority circuit problem is shown below. This solution is also the minimum logic block solution and the minimum signal delay solution.

*Karnaugh Map*

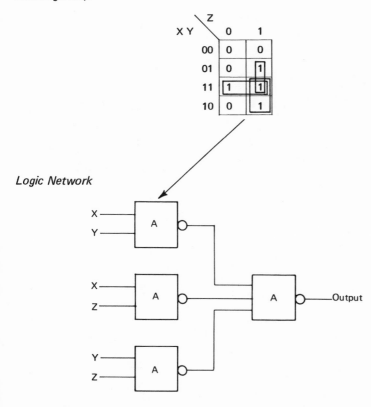

*Logic Network*

*Operation and Application*

The output of this circuit will always agree with the majority of the inputs, and for this reason it is sometimes called a *voter* circuit.

For very reliable switch application, a system called TMR (triple modular redundancy) is often used. With this system all functions are tripled and voters are used to determine the correct outputs.

## Dissent Circuit

The following network is a minimum logic block solution to the problem of producing a signal when one out of three lines disagree on up or down levels. The assumption here is that input complements are not available. A sum of products solution to this problem requires seven blocks, including three inverters. This network, like a sum of products solution, delays the output signal by three logic blocks.

*Karnaugh Mappings*

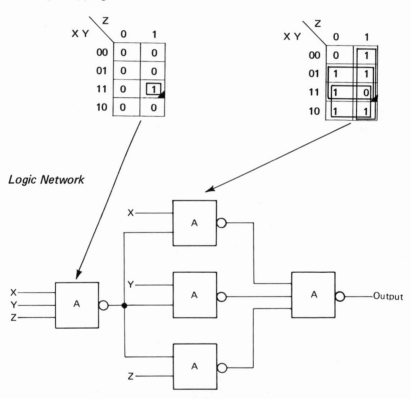

*Logic Network*

Output = $\overline{X}\,\overline{Y}\,Z + \overline{X}\,Y\,\overline{Z} + \overline{X}\,Y\,Z + X\,Y\,\overline{Z} + X\,\overline{Y}\,\overline{Z} + X\,\overline{Y}\,Z$

*Operation*

The output of this network will be up if the three inputs do not agree.

*Application*

This network is useful when working with triple modular redundancy, for it will produce an output signal when the three match signals differ.

### Odd Circuit—Three Inputs No. 1 (Exclusive-OR Tree)

A three input odd circuit is shown below. The output of this network will be up when one or three inputs of a three input set are up. This is a sum of products solution, which means it has minimal signal delay. The network requires five logic blocks when input complements are available and eight blocks if input complements are not available.

*Karnaugh Map*

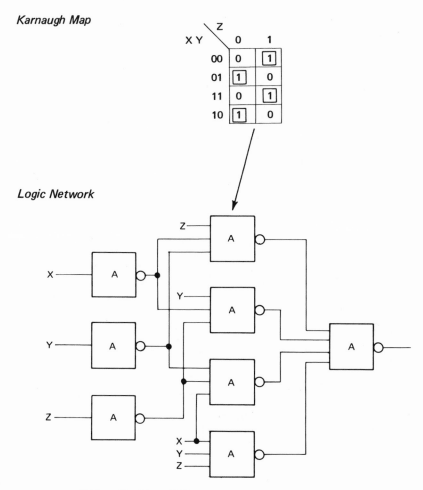

*Logic Network*

Output = $\overline{X}\,\overline{Y}\,Z + \overline{X}\,Y\,\overline{Z} + X\,\overline{Y}\,\overline{Z} + X\,Y\,Z$

*Operation*

The output will be up when an odd number (one or three) of inputs are up.

## Odd Circuit—Three Inputs No. 2

Assuming input complements are not available, the following is a minimal logic block solution to the odd circuit problem. This network requires only seven logic blocks, while the sum of products solution requires eight under the same available input conditions. Input signal Y is delayed by three blocks, while signals X and Z are delayed by four blocks each. The sum of products solution gave a uniform delay of three blocks for all inputs.

*Karnaugh Mappings*

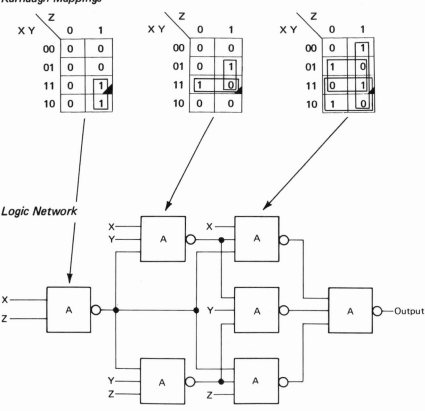

Output $= \bar{X}\,\bar{Y}\,Z + \bar{X}\,Y\,\bar{Z} + X\,\bar{Y}\,\bar{Z} + X\,Y\,Z$

**Odd Circuit—Three Inputs No. 3**

The following three input odd circuit was constructed from two Exclusive-OR networks. The solution requires one additional logic block over the minimal solution, but it has an interesting characteristic. The total number of logic block inputs is 16, while the total number of block inputs for the minimal solution is 20, assuming input complements are not available.

*Logic Network*

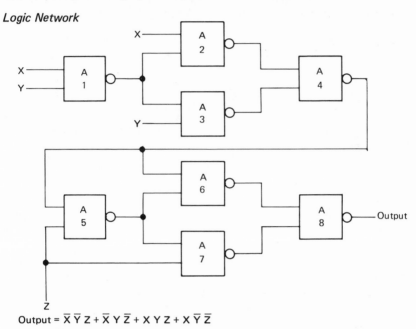

$$\text{Output} = \overline{X}\,\overline{Y}\,Z + \overline{X}\,Y\,\overline{Z} + X\,Y\,Z + X\,\overline{Y}\,\overline{Z}$$

*Operation and Expansion*

The output of NAND 4 produces the Exclusive-OR function of X and Y which is $X\,\overline{Y} + \overline{X}\,Y$. This function is then Exclusive-ORed with Z and the desired output function is obtained. A four input odd circuit may be obtained by connecting another Exclusive-OR network to input Z.

**Even Circuit—Three Inputs**

Assuming no input complements are available, the following network is a minimal logic block solution to the even circuit problem. This network requires seven logic blocks, while a sum of products solution requires eight. Input Z is delayed by three blocks, while inputs X and Y are delayed by five blocks.

*Karnaugh Mappings*

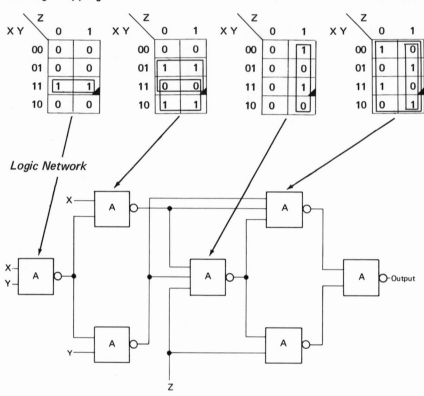

Output = $\overline{X}\,\overline{Y}\,\overline{Z} + \overline{X}\,Y\,Z + X\,Y\,\overline{Z} + X\,\overline{Y}\,Z$

### Odd Circuit—Four Inputs (Also Even and Odd Networks of More Than Four Inputs)

The following network is a minimum logic block solution to the odd circuit problem under the assumption that input complements are not available. This network requires 10 logic blocks, while a sum of products solution requires 13. In the four variable case, the output signal is delayed by five blocks, and seven blocks for the eight variable case.

*Karnaugh Map*

| W X \ Y Z | 00 | 01 | 11 | 10 |
|-----------|----|----|----|----|
| 00        | 0  | 1  | 0  | 1  |
| 01        | 1  | 0  | 1  | 0  |
| 11        | 0  | 1  | 0  | 1  |
| 10        | 1  | 0  | 1  | 0  |

*Logic Network*

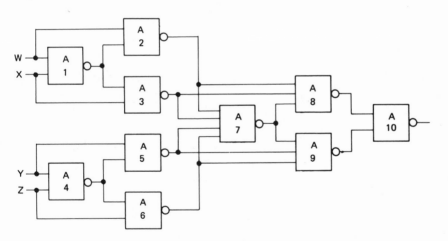

$$\text{Output} = W\,\overline{X}\,\overline{Y}\,\overline{Z} + \overline{W}\,X\,\overline{Y}\,\overline{Z} + \overline{W}\,\overline{X}\,Y\,\overline{Z} + \overline{W}\,\overline{X}\,\overline{Y}\,Z + W\,X\,Y\,\overline{Z} + W\,X\,\overline{Y}\,Z + W\,\overline{X}\,Y\,Z$$
$$+ \overline{W}\,X\,Y\,Z$$

## Design Technique

With some difficulty the design of this network could be explained on the four variable map, but in actual practice the circuit was derived by an obvious extension of the three input even circuit. The Z input of the even circuit was replaced with a network that fed the Exclusive-OR complement function at that point.

## Application and Extension

This network is particularly useful for parity checking or for parity bit generation. When the network is used in either of these applications, it may be necessary to extend the number of inputs. For this extensioning remove block 10 and relabel blocks 7, 8, and 9 as 1, 2, and 3, respectively. Using these three blocks as a starting point, replicate the entire network. This will produce a six input even circuit. If the bottom section of the appearing pyramid is replicated, then an eight input odd circuit is obtained.

## Decoder—Two Variables

This network is the minimum logic block solution to the two variable decoder problem. The output signal is delayed by three logic blocks.

### *Operation*

One and only one of the outputs will be down for each of the four possible input conditions.

### *Karnaugh Mappings*

These maps relate to the output functions only and not to the design procedure which was trial and error.

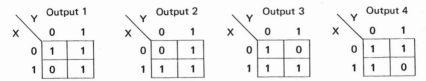

| Output 1 | | |
|---|---|---|
| X \ Y | 0 | 1 |
| 0 | 1 | 1 |
| 1 | 0 | 1 |

| Output 2 | | |
|---|---|---|
| X \ Y | 0 | 1 |
| 0 | 0 | 1 |
| 1 | 1 | 1 |

| Output 3 | | |
|---|---|---|
| X \ Y | 0 | 1 |
| 0 | 1 | 0 |
| 1 | 1 | 1 |

| Output 4 | | |
|---|---|---|
| X \ Y | 0 | 1 |
| 0 | 1 | 1 |
| 1 | 1 | 0 |

### *Logic Network*

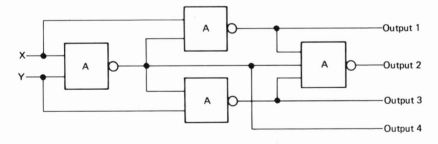

## Decoder—Three Variables

This network is the minimum logic block solution to the three variable decoder problem, assuming input complements are not available. The input signals are delayed by four logic blocks.

*(See Logic Network diagram on following page.)*

*Operation*

One and only one of the outputs will be down for each of the eight possible input conditions.

*Output Expressions*

$$1 = \overline{X\ Y\ Z}$$
$$2 = \overline{X\ Y\ \overline{Z}}$$
$$3 = \overline{X\ \overline{Y}\ Z}$$
$$4 = \overline{\overline{X}\ Y\ Z}$$

$$5 = \overline{\overline{X}\ Y\ \overline{Z}}$$
$$6 = \overline{X\ \overline{Y}\ \overline{Z}}$$
$$7 = \overline{\overline{X}\ \overline{Y}\ Z}$$
$$8 = \overline{\overline{X}\ \overline{Y}\ \overline{Z}}$$

*Restrictions*

The sum of products solution to this problem requires 11 logic blocks, but the signals would be delayed by only three blocks vs. the four in this network.

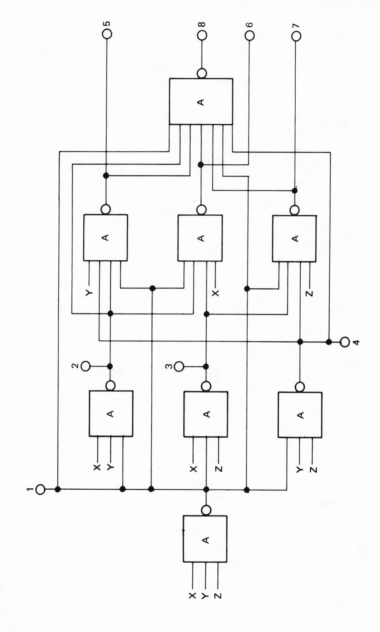

*Logic Network*

**Adder—Half**

The following multioutput network is the minimum logic block solution to the half adder problem, assuming input complements are not available. The sum output signal is delayed by three logic blocks, but of more importance is the fact that the carry output is delayed by only two blocks.

*Karnaugh Mappings*

Sum = $X \overline{Y} + \overline{X} Y$
Carry = $X Y$

*Logic Network*

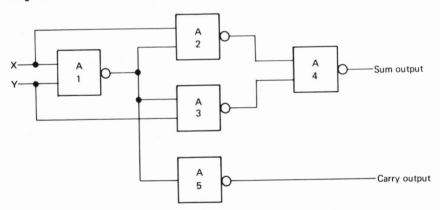

*Design Technique*

From the two variable Karnaugh maps it is obvious that the term $X Y$ is needed as an inhibit term for the sum output, and it is the sole prime implicant for the carry output. Thus the term $X Y$ (block 1) is used in both functions.

*Application*

This network is an important element of many binary adders. A half adder will effectively add two binary positions together, producing sum and carry outputs. The carry output is connected as an input to the next higher position, which must have the ability to add three binary inputs together. The network for this higher order position is called a full adder.

**Adder—Full No. 1**

The following network is the sum of products solution to the full adder problem. The carry signal is delayed by two blocks, as is the sum signal.

*Karnaugh Mappings*

*Logic Network*

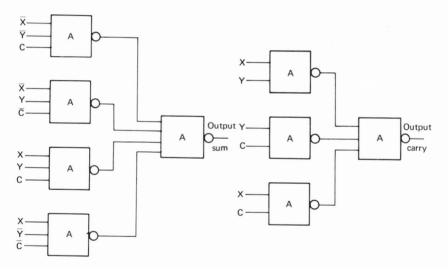

*Application*

This double network has the ability to add together three binary digits. Many of these networks are connected in parallel to form a complete parallel binary adder. The carry output of the low order position is connected as an input to the next higher position, and so on.

*Restriction*

This particular network is of little practical value, since it requires 12 logic blocks if no input complements are available and nine if all are available. In addition, it is not particularly fast or rich in useful subfunctions.

**Adder—Full No. 2**

The following network is a minimum logic block solution for condition of no input complements available (eight blocks).

The most critical delay path in any adder is from carry-in to carry-out. This delay will accumulate as the carry signal is rippled down a parallel adder. A full adder that delays this ripple signal more than two blocks is of little value in most applications. This circuit delays the carry signal by two blocks.

*(See Logic Network diagram on following page.)*

*Restrictions*

While this network has some desirable characteristics it has the following limitations:

1. Block 1 must drive five other blocks.
2. Subfunctions $X + Y$ or $X \overline{Y} + \overline{X} Y$ are not readily available. Either of these subfunctions is referred to as a *propagate term.* As will be shown later, a propagate signal will be needed to design parallel binary adders of any reasonable speed capability.

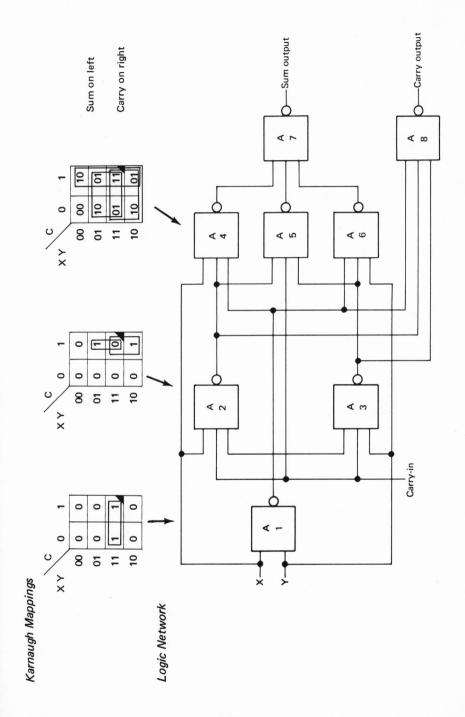

*Karnaugh Mappings*

Sum on left

Carry on right

*Logic Network*

63

## Adder—Full No. 3 (Sum Output Is Complemented)

Minimum solution, assuming input complements are not available and the complement of the sum is permissible as an output.

Signal delay from carry-in to carry-out is two blocks. From the logic network it appears that the carry-in signal is delayed by three blocks (4, 6, and 8), but by examining the related Boolean functions it can be shown that a carry signal passing through block 4 will not change the output of block 6.

*Karnaugh Mappings*

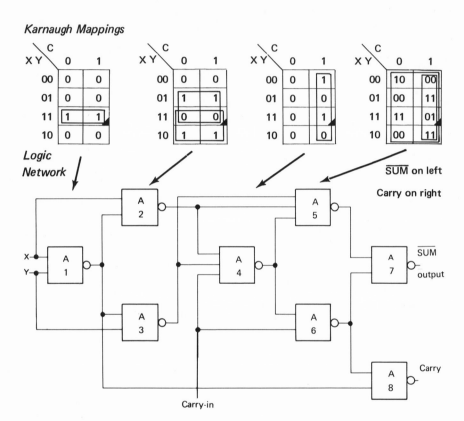

*Logic Network*

SUM on left

Carry on right

Carry-in

*Application*

This full adder is ideal for inexpensive applications. The propagate term explained in full adder No. 2 is not immediately available from this network, but it can be obtained by connecting the output of blocks 2 and 3 to a ninth NAND block.

**Adder—Full No. 4**

A full adder designed from two Exclusive-OR networks.

Total number of NAND blocks is one more than required in a minimal solution (nine vs. eight blocks).

Carry signal delayed by two blocks.

*Karnaugh Map*

|  X Y  | C  0 | 1  |
|-------|------|-----|
| 00    | 00   | 10  |
| 01    | 10   | 01  |
| 11    | 01   | 11  |
| 10    | 10   | 01  |

Sum on left
Carry on right

(*See Logic Network diagram on following page.*)

The output of block 4 is $X \overline{Y} + \overline{X} Y$, a desirable subfunction as previously explained. Although this network requires nine NAND blocks, each of the blocks has only two inputs. This network has fewer total inputs than full adder No. 2 which has a minimum number of blocks.

*Logic Network*

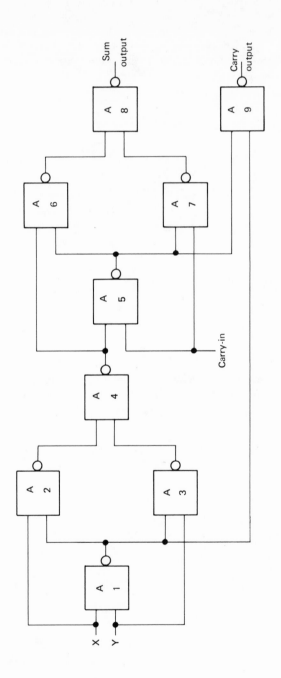

**Subtractor—Full**

Nonminimal solution constructed from two Exclusive-OR networks.

An interesting relationship exists between this full subtractor and full adder No. 4, the difference between the two networks being the connections to block 9. For the full adder network block 9 is connected to blocks 1 and 5, and in this network it is connected to blocks 3 and 7.

*(See Logic Network diagram on following page.)*

*Operation*

The full subtractor performs the same function for subtraction that the full adder performs for addition. Input Y is subtracted from X and input "borrow" is also subtracted from X. The borrow output should be connected to the borrow input of the next higher position in the design of a parallel subtractor. In a parallel subtractor the difference will appear in two's complement form if the subtrahend is greater than the minuend. However, the one's complement form of negative numbers can be obtained under the same conditions by connecting the high order borrow output to the low order borrow input.

*Karnaugh Map*

B

X Y | 0 | 1

00 | 00 | 11
01 | 11 | 01
11 | 00 | 11
10 | 10 | 00

Difference on left
Borrow on right

*Logic Network*

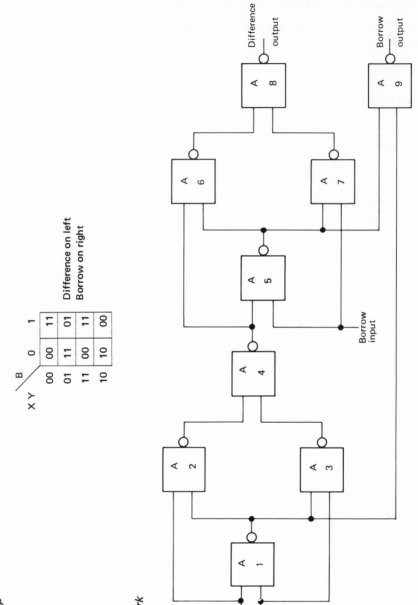

X
Y

A 1

A 2
A 3

A 4

A 5

Borrow input

A 6
A 7

A 8

Difference output

A 9

Borrow output

68

**Adder Speed-Up Circuit No. 1**

*Basic Principle*

The speed of a parallel binary adder is determined mainly by the delay of the carry signal as it propagates down the series of full adders. This delay may be substantial for it is accumulative at the rate of two logic blocks per bit position. The following circuit is designed to avoid much of this delay by enabling the carry ripple to bridge or pass over those full adder positions that are appropriately conditioned by input signals. The network depends upon a propagate signal from each of the full adders. The propagate line of each full adder should be conditioned whenever the X and Y input lines of that position are such that a carry-in signal will generate a carry-out signal. The Boolean expression for the propagate signal is $X + Y$ or $X \overline{Y} + \overline{X} Y$. The added NAND block in the following network is connected as an input to the last NAND block in the carry network of the last full adder position.

*Logic Network*

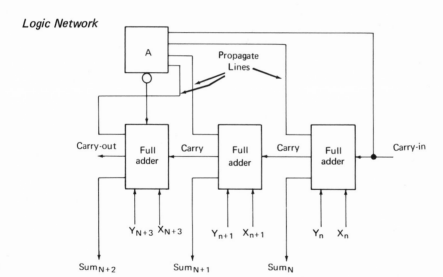

*Operation*

In this example a carry-in signal will be delayed by only two logic blocks instead of six as it ripples past this group.

*Extension*

This bridging of the carry signal may be expanded to any number of full adder positions as long as each bypassed stage furnishes a propagate signal to the added NAND block.

*Restriction*

This speed-up network does nothing for a carry signal generated within the bridged group.  Speed-up circuit No. 2 shows how an additional NAND is used to resolve this problem.

## Adder Speed-Up Circuit No. 2

This network is identical to speed-up circuit No. 1 except that a second level of bridging has been included to decrease signal delay.

*Logic Network*

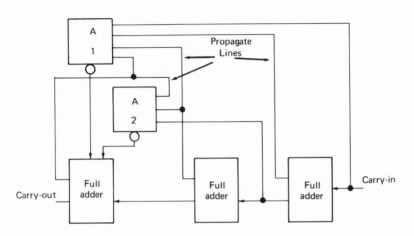

*Operation*

The function of block 1 has been explained in the preceding circuit. The output of block 2 is connected as an input to the last NAND block in the carry network of the indicated full adder; blocks 1 and 2 connect to the same block. The function of block 2 is to speed up a carry signal that is started in the rightmost full adder. This signal would normally pass through two blocks in the middle full adder and two more blocks in the leftmost adder, a total delay in this case of four blocks. But by adding block No. 2 to the network, this carry signal will be delayed by only two blocks, block No. 2 and the last NAND in the leftmost full adder.

*General*

The speed-up principle shown in this and the preceding network can be applied and reapplied. The bridges can be as long as desired, and any number of levels may be used. The number of levels and the length of each level are functions of many variables. Number of inputs allowed on each block, number of outputs that can be driven, and even packaging and wiring rules are factors in this problem. For this reason no general rules can be given, but it is not uncommon for bridging to require as many logic blocks as are required by the basic full adders themselves.

## NAND SEQUENTIAL NETWORKS

### Flip-Flop Latch (Flip-Flop, Set/Reset)

The following network is the basic storage element in the NAND logic family. It has two stable states when both inputs are up and can be placed in either of these states by lowering the appropriate input. The network is said to be *set* when output A = 1 and output B = 0. This output condition is arrived at by lowering the normally up set line. Likewise, the network may be reset by lowering the *reset* input. When the network is reset, output A = 0 and output B =1. The returning of a lowered input to its upper value has no effect on the output.

A complete chart is used to show relationship of outputs.

*Flow Chart*

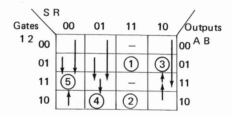

*Logic Network*

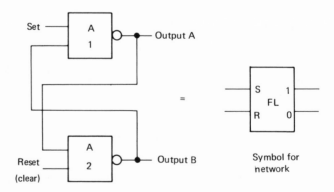

*Operational Notes*

1. Outputs A and B are complements of each other except when both inputs are down, in which case both outputs are 1.
2. If both inputs are lowered at the same time, the final output values will be determined by the input that is held at the 0 value the longest.
3. Theoretically, the network can be made to oscillate by lowering both inputs and then raising them together. This will not happen in actual practice, but the final state of the network is unpredictable ( ① or ② ) under this double input change.

*Additional Inputs*

Additional set lines may be added to block 1 just as additional reset lines may be added to block 2. All such inputs should be held at the 1 level when not in use. The lowering of any input will, respectively, set and reset the network.

**Inverter, Nonoverlapping**

This network generates a true and an inverted signal which are nonoverlapping in the down level; both outputs will never be down at the same time.

Many ring networks and shift registers require driving lines that are nonoverlapping.

*Input-Output Timing Chart*

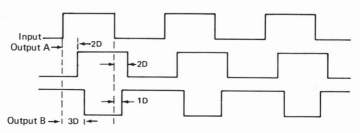

D = Delay through one NAND gate

*Flow Chart*

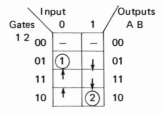

*Logic Network*

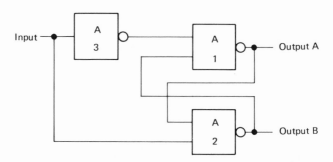

*Operational Notes*

As the input changes from 0 to 1 and back to 0, the network moves from ① to ② and back to ①. It is important to note that the network never moves directly to the next stable state but first passes through an unstable state where the output lines are both up. This technique prevents both outputs from ever being down at the same time.

**Contact Bounce Eliminator**

Connecting mechanical switches to high speed electronic logic can introduce problems as a result of the bouncing of the mechanical contacts. In the closing of a switch the contacts may touch and bounce open several times before the contact is made and held. This bouncing will generate a string of pulses rather than a voltage shift as is usually desired. Fortunately, most switches are available with a double throw contact. Switches of this type have a center strap (common strap) that will bounce when it is making or breaking contact, but the normally open and normally closed contacts are usually so far apart that the center strap will not bounce between the two. Once the center strap is able to touch the contact it is closing on, it is impossible for it to recontact the strap that has just opened. Under this condition it is possible to design a logic network that will eliminate any appearance of the bouncing.

*Logic Network*

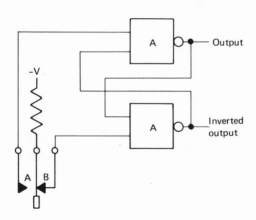

*Flow Chart (Partial)*

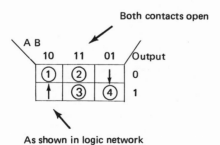

As shown in logic network

## Flip-Flop, Gated No. 1

A flip-flop with one or more inputs where each input is under control of a separate gate line. The set signal is delayed by two blocks before appearing on output A.

*Flow Chart (Partial)*

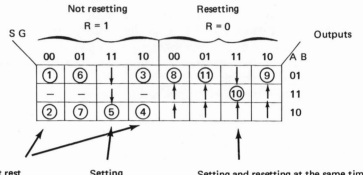

*Logic Network*

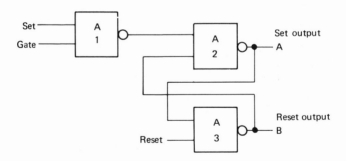

## Operational Notes

1. Reset line is normally up and the gate line is normally down.
2. Network is independent of all changes on the set line as long as the gate line is down.
3. Outputs A and B are complements of each other except when attempting to set and reset simultaneously. When this input condition occurs, the network will move to stable state ⑩ and both outputs will go up.
4. Data on the set line is delayed by two blocks before appearing on output A, three blocks for output B.
5. The gate line must be up for the delay time of two blocks to insure the setting of the latch. The reset line must be down for the time delay of two blocks to insure resetting.
6. For high speed operation the gate and reset signals are usually overlapped, as shown below.

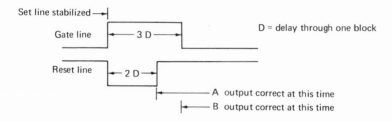

## Additional Inputs

Additional gated set lines may be added by duplicating block 1. The network will be independent of all set lines except when a related gate line is raised. Only one gate line should be raised at a time.

**Flip-Flop, Gated No. 2**

The following network has a high speed gated set input. Set information is delayed by only one gate. The set output signal appears on two wires that must be ANDed together. This design approach is based on the assumption that the given network connects to a NAND gate in a following network. It is further assumed that this receiving NAND can tolerate one more input and therefore can perform the ANDing function without adding to the signal delay.

When the network is not being changed, the reset line and the gate line are both up.

*Flow Chart*

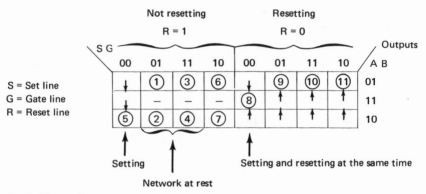

*Logic Network*

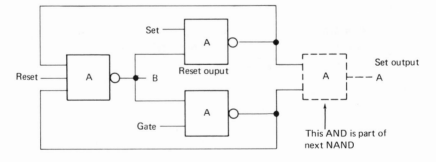

*Operational Notes*

1. As stated, when the flip-flop is not being changed the reset line and the gate lines are up.
2. The set output will be up when the network is set and down when the network is reset.
3. In order to set the network, the set line must be *down* when the gate line goes down. Output A, the set output, will respond in the time it takes the signal to pass through one logic gate.
4. To reset the network the reset line is lowered. This will lower output A and raise output B.
5. For high speed operations the following gate and reset timings should be used:

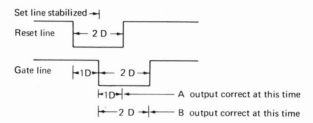

6. Notice that three units of delay are required to take the network from any state to any other state. But again, it should be noted that the set data is delayed by only one unit of delay.

### Flip-Flop, Automatic Reset Type No. 1

The gated flip-flop shown below requires no reset line. The flip-flop is conditioned to the value of the data line when the gate line is raised. When the gate line is down the flip-flop is insensitive to the data line.

*Input-Output Timing Chart*

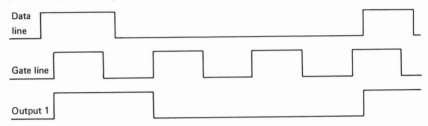

*Flow Chart*

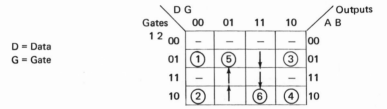

*Logic Network*

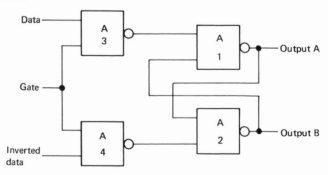

*Operational Notes*

1. The network requires the availability of a data line and its complement.
2. The two output lines are complements of each other except when the flip-flop is changing state, in which case they are both at the 1 level.

*Additional Inputs*

Separate set and reset lines may be brought into logic gates 1 and 2, respectively. These inputs must be up except when they are being used to set or reset the flip-flop.

Additional gated data lines may be added to the network by duplicating gates 3 and 4 and connecting them to gates 1 and 2, respectively. In this case the raising of either gate line will cause the data on the respective data line to be entered into the flip-flop.

**Flip-Flop, Automatic Reset Type No. 2**

This network is similar to the previous flip-flop and it has the advantage of not requiring the inverted data line. However, this network is slower in responding to one type of input change. If the data line is changed while the gate line is up, the outputs may not stabilize until the data signal has passed through four gates. In the previous network this condition resulted in a three gate delay.

*Input-Output Timing Chart*

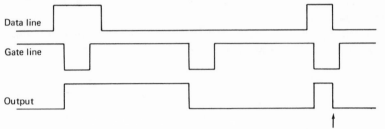

This is the delay condition described

*Flow Chart*

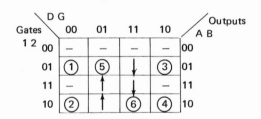

*Logic Network*

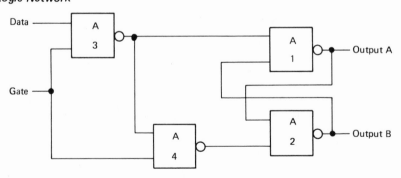

*Design Note*

Gate 3 performs two functions in this network: it gates the set input to the flip-flop (gates 1 and 2) and provides the inverted data line for gate 4.

*Operational Notes*

1. The outputs are normally complements of each other except when the network is changing state, in which case the outputs will both be at the 1 level.
2. When the gate line is raised, the flip-flop will be conditioned to the value of the data line.
3. When the gate line is down, the flip-flop is insensitive to changes on the data line.

*Additional Inputs*

Separate set and reset lines may be connected to gates 1 and 2, respectively. These lines must be up when not in use.

*Double Gating*

This particular network is well suited to applications where a timing signal and a gating signal are used. In these applications the gating signal is used to determine which input will feed the flip-flop, and the timing signal is used to determine solely the time at which the data enters the flip-flop. With this system the proper gate line must be raised before the timing line is raised and must be maintained throughout the period of the timing signal.

*Logic Network for Double Gating*

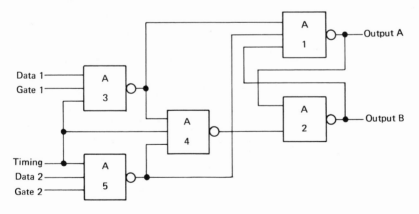

### Flip-Flop, Automatic Reset Type No. 3

This network, like the other automatic reset type flip-flops, requires no reset line. This network is conditioned by the fall of the gate line rather than the rise of the gate line. When the gate line is up, the flip-flop is insensitive to changes on the data line.

*Input-Output Timing Chart*

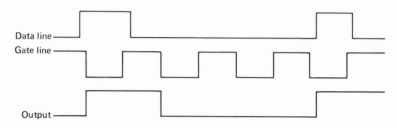

*Flow Chart*

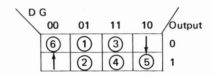

*Logic Network*

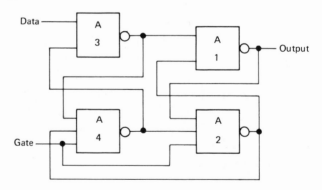

*Operational Notes*

1. The output signal is delayed by three gates from the time the gate line is lowered.
2. A complemented output signal is not available. The output from gate 2 can, in some applications, fill the need for a complemented output. The expanded flow table shown below fully explains the output of gate 2 and how it relates to gate 1.

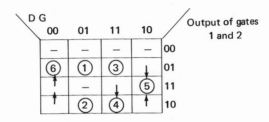

*Additional Inputs*

A separate reset line may be connected to gates 2 and 3. This input should be up when not in use.

*Design Note*

This network has been conservatively designed and therefore has the appearance of being overly complex. In this network only one gate will change its output at a time; there are no race conditions between two or more gates. Networks of this type are usually difficult to design and are discovered by trial and error.

This same circuit without the race removing lines is shown in the following network.

## Flip-Flop, Automatic Reset Type No. 4

The following is by far the most popular of all the automatic reset type flip-flops. When used in registers where a series of flip-flops are all gated at the same time, this network will require only three gates in place of the usual four.

### Timing Chart

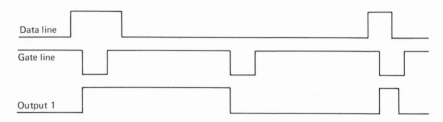

### Flow Chart

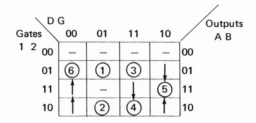

### Logic Network

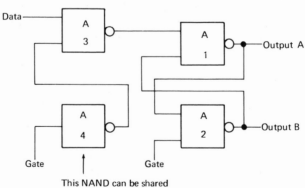

This NAND can be shared
with all positions of a register

*Operational Notes*

1. The network will be conditioned to the data line when the gate line is lowered. When the gate line is up, the network is insensitive to the data line.
2. Output B is not at all times the complement of the A output line. See flow chart for detailed operation of output B.
3. There is a theoretical race condition in this network when the gate line is being raised. If the gate signal into NAND 2 is delayed and the gate signal passing through NAND 4 is not, then it is conceivable that the flip-flop will fail to set. This gate condition can easily be avoided, but care should be exercised in generating and powering up these gate lines.

## Flip-Flop, Three State

This network is the natural extension of the two state flip-flop. As with the two state flip-flop, all NAND gates feed all other NAND gates.

This network has three stable states when all inputs are up.

*Partial Flow Chart*

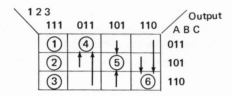

*Logic Network*

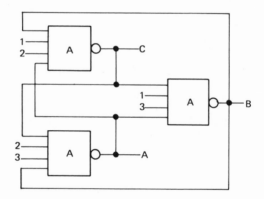

*Operational Notes*

1. When the network is not being changed all inputs should be up. Under this input condition one and only one of the outputs will be down.
2. The flip-flop may be set to any of its three states by lowering a corresponding input set line. The flip-flop will then remain in this state as the input line is raised.
3. When two or all three inputs are down at the same time, all outputs will be up and the state of the flip-flop will be determined by the input which is raised last.

**Flip-Flop, Four State**

This network, like the three state flip-flop, has all gates connected to all other gates.

At first glance this network appears overly complex, since two of the well-known two state flip-flops will when operated together produce four states. The following network, however, is unique in that one and only one of the four outputs is shown for each of the possible four states. The output is said to be coded in a 1 out of n code. To obtain output lines of this type from two flip-flops, each having two states, would require the addition of four AND gates.

*Logic Network*

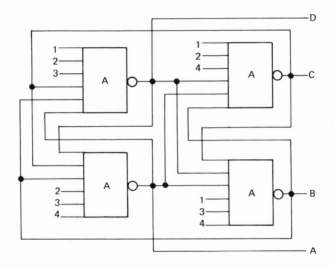

*Operational Notes*

1. When the network is not being changed, all four inputs are normally up. Under this input condition, the network has four stable states.
2. One and only one of the output lines will be down when all four inputs are up.
3. The network may be moved from one stable state to another by lowering then raising the appropriate input.

| Input lowered | Output condition |
|---|---|
| | A  B  C  D |
| 1 | 0  1  1  1 |
| 2 | 1  0  1  1 |
| 3 | 1  1  0  1 |
| 4 | 1  1  1  0 |

## The Five State Flip-Flop

Flip-flops with five or more states can be constructed by extending the philosophy that was used to obtain the three and four state flip-flops. For a five state flip-flop, interconnect five NAND gates so that each gate feeds all other gates. Four of the five input lines should connect to each of the NAND gates in a unique pattern.

## Flip-Flop, T Type No. 1

A flip-flop whose output changes each time the input line is raised. Lowering of the input line has no effect on the output. Flip-flops of this type are used in the construction of binary counters.

### Flow Chart

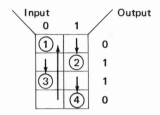

### Timing Chart

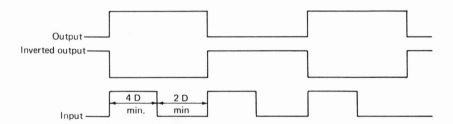

### Logic Network

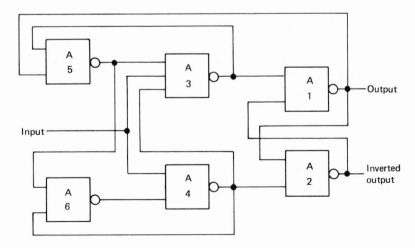

## Operational Notes

T type flip-flops are by far the most interesting of the normally used sequential networks. For this reason a complete timing chart of all gates in the network is shown below. The output signals from these internal gates often have useful applications.

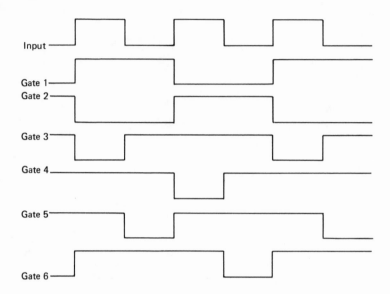

Notice that gates 3, 5, 4, and 6 successively produce negative pulses as the network is moved through its stable states.

## Additional Inputs

This network can be set to an output of 1 or reset to an output of 0 by adding lines to gates 1 and 2, respectively. Lowering the line into gate 1 will set the output to 1, while lowering the line into gate 2 will reset the output to 0. These inputs will be effective only when the input is down. This is an important note and should not be overlooked.

*General Information*

As mentioned, flip-flops of the T type are often used in the construction of binary counters. The design of binary counters is covered later in this manual, but to facilitate their design an additional output is of value. This output should provide a signal indicating that the flip-flop has just changed from 1 to 0. This output should be up only when the input to the flip-flop is up, as shown in the following timing chart.

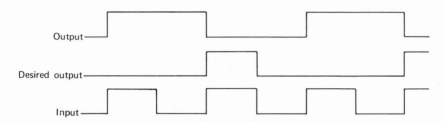

This desired output can be obtained by ANDing together the outputs of gates 2 and 6. (For application of this signal, see binary counter No. 2.)

## Flip-Flop, T Type No. 2

A T type flip-flop that changes state on the raise of the input line. This flip-flop is preferred over the previous flip-flop for the following reasons:

1. Network can operate at its maximum rep-rate with a symmetrical input pulse pattern (see timing chart).
2. Network can be set or reset with the input line up or down.

*Flow Chart*

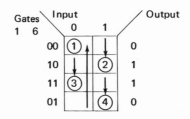

*Timing Chart*

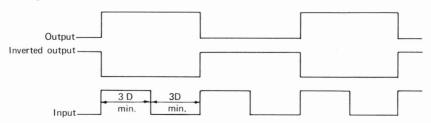

*Logic Network*

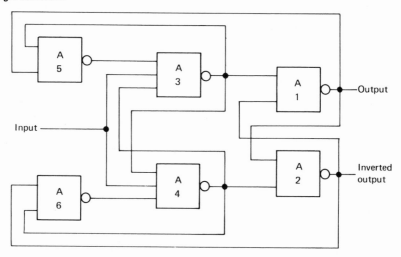

## Operational Notes

The output signal and the inverted output signal are never more than three units of delay behind the input signal, a unit of delay being the delay through any single gate.

## Additional Inputs

This flip-flop may be set to an output condition of 1 or reset to an output condition of 0 by adding lines as indicated in the following table.

| Condition of input when setting or resetting is to occur | Desired output condition | Add an input line to the following gates |
|---|---|---|
| Input line down | Output = 1<br>Output = 0 | Gate 1<br>Gate 2 |
| Input line up | Output = 1<br>Output = 0 | Gates 4 and 5<br>Gate 6 |
| Input unknown | Output = 1<br>Output = 0 | Gates 1, 4, and 5<br>Gates 2 and 6 |

All added lines must be maintained at the 1 level when not being used.

## General Information

Since this network is used as a building block for more complex networks, it is advisable to examine a timing chart which covers the timing of all six gates.

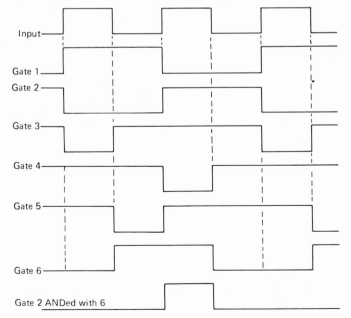

## Flip-Flop, T Type No. 3

A flip-flop whose output changes each time the input line is lowered. Raising the input has no effect on the output.

### Flow Chart

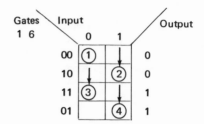

### Timing Chart

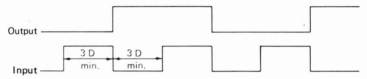

### Logic Network

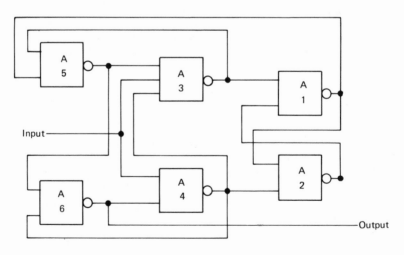

*Operational Notes*

This is the same network as used in the previous T type flip-flop except the output is taken from gate 6 instead of gate 1.

*Additional Inputs*

Use the following table for adding lines which will set the output of the flip-flop to 1 or reset the output to 0.

| Condition of input when setting or resetting is to occur | Desired output condition | Add an input line to the following gates |
|---|---|---|
| Input line down | Output = 1<br>Output = 0 | Gate 1<br>Gate 2 |
| Input line up | Output = 1<br>Output = 0 | Gate 6<br>Gates 4 and 5 |
| Input unknown | Output = 1<br>Output = 0 | Gates 1 and 6<br>Gates 2, 4, and 5 |

All added lines must be maintained at the 1 level when not being used.

### Flip-Flop, J-K type

Like the simple set/reset flip-flop, this network has two inputs. Raising the J input will set the output to 1, while raising the K input will reset the output to 0. But unlike the set/reset flip-flop, the raising of both inputs will cause the output to change. If the output is 1 when both inputs are down, then the raising of both inputs will cause the output to change to 0; and likewise if the network started with an output of 1, it would change to 0 as the inputs are raised.

*Flow Chart*

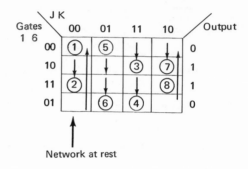

Network at rest

*Logic Network*

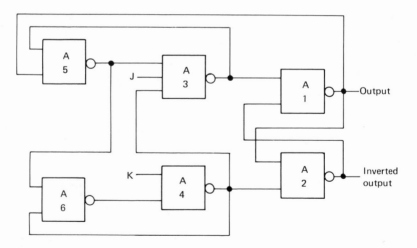

*Operational Notes (Taken from Flow Chart)*

As stated, the raising of both inputs at the same time will cause the output to change. If the network is at state ① it will move to ③ , or if at state ② it will move to state ④ . The simultaneous lowering of both inputs has no effect on the output. If at state ③ the network will move to ② or if at ④ it will move to ① as the inputs are lowered.

From the flow chart it should be observed that both inputs do not have to be raised at precisely the same time to cause the output to change. It is only required that both inputs be up at the same time to effect the change. As the network moves from ① to ③ it may stop at ⑤ or ⑦ if one of the inputs is slow in rising. If the network starts at ② it may stop at ⑥ or ⑧ as it moves to ④. However, this same laxity in timing does not exist in the lowering of both inputs. The lowering of both inputs cannot be separated timewise by more than one gate delay. This hazard can be seen as the network moves from ③ to ⑤ . If K is lowered first, the network moves to ⑦ which is acceptable since it then will move to ② as J is lowered. But what if J is lowered before K? The network will move from ③ to the next left column and will start to step to state ⑥ . The K input must now be lowered before the network arrives at ⑥ or the final state will be ① instead of the correct ② . The same type of timing problem exists as the network moves from ④ to ① , but in this case J cannot be later than one unit of time from K.

## Flip-Flop, Set Dominant

A set dominant flip-flop is a two state network having a set and a reset input. It differs from a conventional set/reset flip-flop in that an attempt to simultaneously set and reset will result in the setting of the network. This particular network is at rest when both inputs are up, and the lowering of an input indicates a setting or resetting operation.

*Flow Chart*

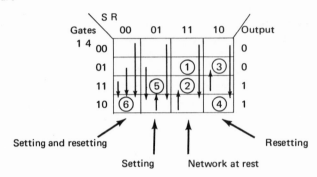

Setting and resetting

Setting     Network at rest

Resetting

*Logic Network*

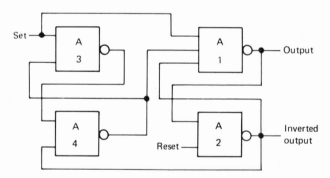

*Operational Note*

Notice from the flow chart that once column 00 (simultaneous setting and resetting) has been entered, the order of returning the inputs to their up levels is unimportant. If the set line is raised first, or the reset is raised first, or for that matter if they are raised at the same time, the output remains at 1.

*Additional Inputs*

If an additional set line is added to gates 1 and 3, then the lowering of either set line will set the flip-flop. An additional reset line may also be added to gate 2, and the lowering of either reset line will reset the flip-flop.

### Flip-Flop, Post Indicating

A post indicating flip-flop is a two state network having two inputs: a set and a reset. The raising of the set line will set the network to a 1, but the output will not change until the set line is returned to its down level. The clear, or reset, input operates in the same manner. Raising the reset input will reset the network, but the output will not change until this input is lowered.

*Flow Chart*

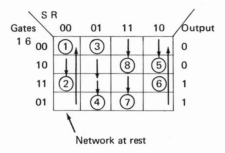

Network at rest

*Logic Network*

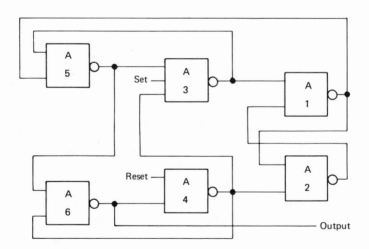

*Operational Notes*

1. Both inputs are normally down.
2. Raising the set line will set the network, but the output will not change until the set line is lowered.
3. Raising the reset line will reset the network, but the output will not change until the reset line is lowered.
4. If both lines are raised the output will not change, but the output value will be determined by the input line that is lowered last (see flow chart).

**Flip-Flop, AC Coupled Type**

This network performs as if the two inputs were AC coupled to the flip-flop, but no capacitors or other elements aside from NAND gates are used. The network responds only to the leading edges (negative going in this case) of incoming pulses.

*Flow Chart*

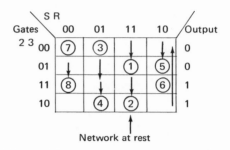

Network at rest

*Logic Network*

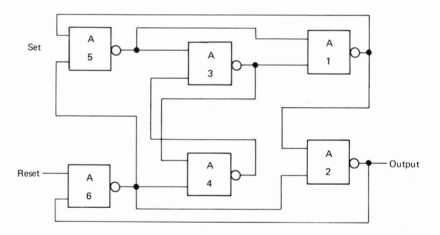

*Operational Notes*

1. Both inputs are normally up. The lowering of the set line will set the output to 1, and lowering the reset line will reset the network to 0.
2. Neither input need be returned to its normally up level to make the other input effective. The flip-flop will be set or reset depending on which of the two inputs was lowered last.
3. The raising of either input will not affect the output.

*Additional Inputs*

Additional set inputs can be connected to gate 5. When this has been done, all set inputs should normally be up; and the lowering of any set line will set the flip-flop. Once a set line has been lowered, the lowering of other set lines will have no effect. All set lines must be returned to the up level before any of them become effective again.

Similar conductions can be stated for additional reset lines connected to gate 6.

## Gated Oscillator No. 1

This network will gate an oscillator signal without shortening or lengthening any of the oscillator pulses. There will be no fractional pulses appearing on the output regardless of the timing of the gate line.

*Input-Output Timing Chart*

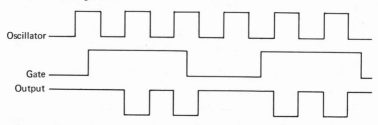

*Flow Chart (Partial)*

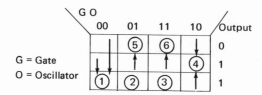

*Logic Network*

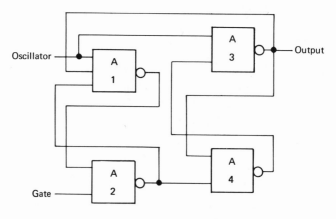

*Operational Notes*

*From the above flow chart:*

1. It is assumed that the oscillator signal is continuously applied to the network. The network will, therefore, move between stable states ① and ② when the gate line is down. The output will remain 1 as the network moves between these two states.

2. For the output to go down, the network must reach stable states ⑤ or ⑥, but the only path to either of these states is through state ④. To get to state ④ the gate line must be up when the oscillator line is down. The network is now ready to move in unison with the oscillator input.

3. As long as the gate line is held up and the starting point was ④, the network will move from ④ to ⑥ to ④, etc.

4. If the gate line is dropped in the middle of an oscillator pulse, the network will move to ⑤ then to ①. State ⑤ has an output of 0 so that the output pulse will be completed.

## Gated Oscillator No. 2

When the gate line is raised, this network will gate one and only one oscillator pulse. The output pulse will always be a full width pulse regardless of the timing of the gate line.

*Input-Output Timing Chart*

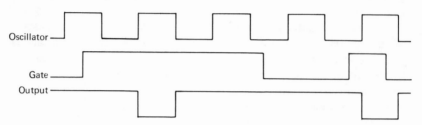

*Flow Chart (Partial)*

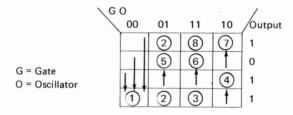

G = Gate
O = Oscillator

*Logic Network*

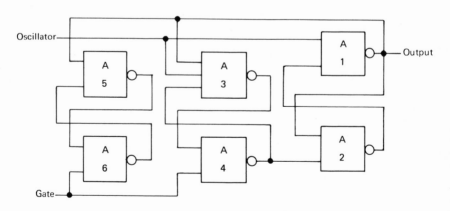

*Operational Notes*

From the above flow chart:

1. It is assumed that the oscillator signal is continuously applied to the network. The network will, therefore, move between stable states ① and ② when the gate line is down. The output will remain as a 1 as the network moves between these 2 states.

2. For the output to go down, the network must reach stable state ⑤ or ⑥ , but the only path to these states is through state ④ . To get to state ④ the gate line must be up when the oscillator line is down. The network is now ready to move with the oscillator signal into state ⑥ , lowering the output. When the oscillator signal drops, the network moves on to ⑦ and is trapped here until the gate line is lowered. Thus the network followed the oscillator for only one cycle.

*General Information*

The first four gates in this network are identical to those of gated oscillator No. 1. A flip-flop, gates 5 and 6, has been added to that basic network to control the gating signal.

**Gated Oscillator No. 3**

This network will selectively perform the functions of gated oscillator No. 1 and gated oscillator No. 2.

When the control line is down and the gate line is raised, this network will gate our a string of inverted oscillator pulses. There will be no partial pulses on the output line regardless of the gate timing.

When the control line is up, the network functions as gated oscillator No. 2. The raising of the gate line will cause the output to follow the inverted oscillator signal for one negative excursion.

*Logic Network*

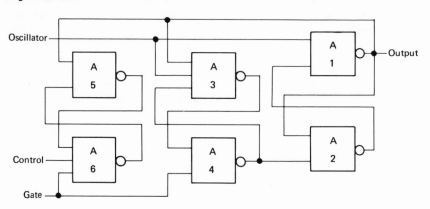

## Sampling Gate, Type 1

This network is sometimes called a picture-taking flip-flop.

The value of the data line is entered into the flip-flop only when the gate line is being raised. The flip-flop is insensitive to the data line when the gate line is down or after it has been raised. In effect, the network takes a picture of the data line when the gate line is moving from 0 to 1.

*Flow Chart*

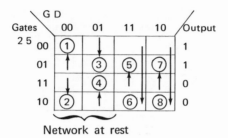

Network at rest

*Logic Network*

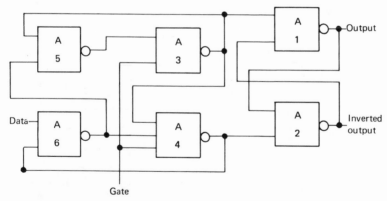

*Operational Notes*

1. Gate line is normally down and the flip-flop is insensitive to all changes on the data line.
2. On the rise of the gate line the value of the data line is entered into the flip-flop. The value in the flip-flop will remain until the gate line is lowered and then raised again.

*Additional Inputs*

Set and reset lines may be added to gates 1 and 2, respectively. These lines should normally be up and will be effective only when the gate line is down.

## Sampling Gate, Type 2

This network provides a similar function as that of sampling gate 1 but requires only four NAND gates. The operational details of this network are somewhat different from those of sampling gate 1 and should be studied carefully. The network in effect takes a picture of the data line on the rise of the gate line but holds that picture only as long as the gate line is up.

*Flow Chart*

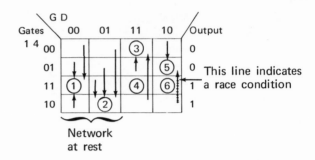

Network
at rest

*Logic Network*

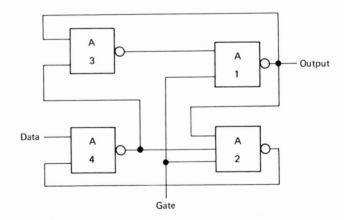

*Operational Notes*

1. Gate line is normally down.
2. The output is inverted. That is, when the network is set the output is down and when reset, the output is up.
3. When the gate line is down, the network is reset (output is 1) and insensitive to the data line.
4. On the rise of the gate line, the value of the data line is locked into the network. The output will assume a value that is the inverse of the data line.
5. When the gate line is up, the network will remain stable at the level the data line was at when the gate line was going up.
6. If the gate line is raised while the data line is coming down, the network encounters a race condition, as shown on the flow chart. The output under this condition may lock on a 1 or a 0. There is little reward in removing this race, since the output will always be in question if the data line is sampled while changing.

**One Shot (Single Shot)**

A *one shot* is a network designed to produce an output pulse of constant width irrespective of the input pulse width. The input and output of this network are normally up.

*Logic Network*

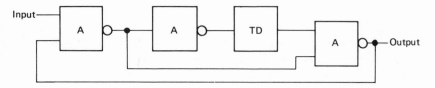

*Timing Chart*

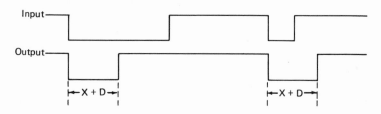

$X$ = the time delay imposed by the delay element
$D$ = the time delay on one NAND GATE

*Restrictions*

After generating an output pulse, the network cannot be recycled until the delay path has been cleared. This requires a time interval of $X + D$ units.

**Pulse Shortener**

The following network has the purpose of shortening an incoming positive pulse. The length of the output pulse cannot exceed a prescribed interval which is determined by a delay unit. The input to this network is normally down, while the output is normally up.

*Logic Network*

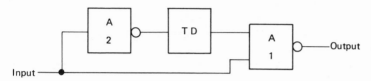

*Timing Chart*

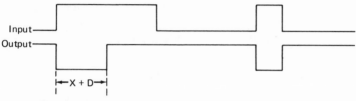

$X$ = the time delay imposed by the delay element
$D$ = the time delay of one NAND gate

*Operational Notes*

The length of the output pulse will be the smaller of the following conditions.

1. Equal to the input pulse.
2. Equal to the time delay of gate 2 plus the delay of the delay element.

*Restrictions*

After generating an output, the network cannot be recycled until the delay path has cleared. The input line must therefore be down for $X + D$ units before being raised again.

The delay unit may be constructed from a delay line or an RC network

**Adder with Latched Output**

This latched adder network is constructed from a full adder and a flip-flop. The particular adder selected is relatively unimportant, but what is important is the method of merging them together. This network uses full adder No. 4.

*(See Logic Network diagram on following page.)*

*Operational Notes*

1. The two binary positions to be added are connected as inputs to X and Y. The carry output of one stage is connected to the carry input of the next higher stage.

2. The gate line is normally down, making the flip-flop (gates 8 and 10) insensitive to the adder.

3. The two binary numbers to be added are presented to the adder; and after sufficient time has elapsed to allow for the completion of the ripple carry, the gate line is raised. This raising of the gate line will set or reset the flip-flop to the value of the sum. When the gate line is lowered, this sum is locked in the flip-flop and the adder is now free to start the next addition.

*Design Notes*

1. Gates 1-9 were generated from the original full adder design.

2. Gate 10 has been connected to gate 8 to form a flip-flop. Gate 11 has been added to the network to reset the flip-flop when the output should be 0. The gate line has been connected to all gates that feed the flip-flop. The outputs of these NANDS will, therefore, be at the 1 level when the gate line is down. All inputs to the flip-flop, excluding the cross connecting lines, will be at the 1 level when the gate line is down, and the flip-flop will be effectively isolated.

*Logic Network*

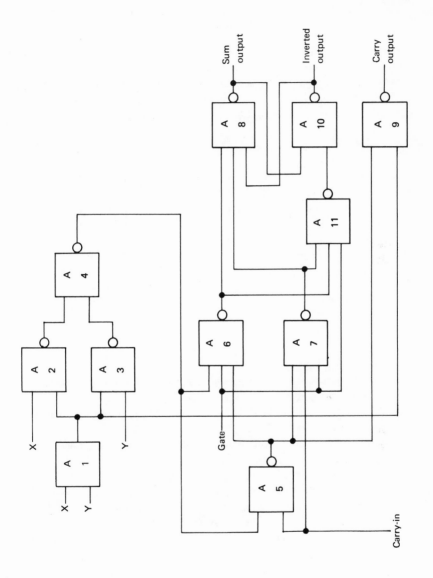

**Binary Counter No. 1**

The following network has been designed to count the number of positive pulses appearing on a line. The count is continually displayed in binary, one output line per T flip-flop. This network is the simplest of all the counters, but it suffers some severe limitations as will be shown. The network is constructed from either T type flip-flop No. 1 or No. 2. These networks change their output state on the raise of the input line.

*Logic Network*

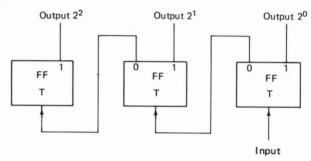

*Operational Notes*

1. The timing chart points out the value of the output lines when the input line is down, but it should be observed that the output lines actually become stable at these values while the input is still up.
2. The input to all flip-flops, except the low order, comes from gate 2 of the next lower order flip-flops.
3. This counter uses a single wire to interconnect adjacent stages. When the input line to the counter is down, these interconnecting lines may be up or down, depending upon the counted value. Some T type flip-flops cannot be set when their input lines are up and therefore cannot be used in this network if setting to values other than 0 is required. Flip-flop T type No. 1 has the above-mentioned restriction.
4. This network is relatively slow, since an input signal will have to ripple through one stage of the counter to the next. The number of stages this input signal will have to ripple depends on the counter, the longest ripple being when the highest position must be changed. The time required for this longest ripple is the time that is used in computing the rep-rate of the counter.

*Timing Chart*

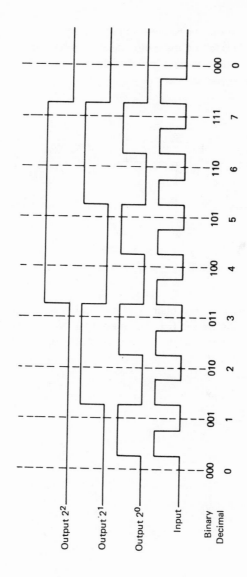

## Design Alternative

The above design may be changed slightly by using flip-flops that change their output states on the fall of the input signal. Flip-flops of this type (flip-flop, T type No. 3) can be used to replace all but the low order stage. When this is done, the flip-flops are connected as shown.

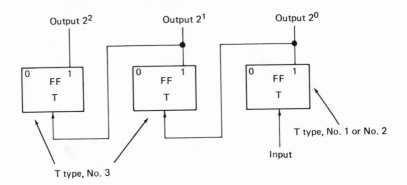

The timing chart for this counter is identical to that of the first counter. The most significant difference between these two counters is the loading of the output line. In the last counter the output lines are required to drive more gates than in the first case.

**Binary Counter No. 2**

This network, like binary counter No. 1, counts in binary the number of positive pulses appearing on the input line. But unlike binary counter No. 1, this network offers no difficulty in being set to any particular binary number before the counting is started. The simplicity in the setting or resetting of this counter is a result of the input signal to each flip-flop being at the down level when the input to the counter is down. Refer to T flip-flop for a timing chart showing the signal that results from ANDing together the outputs of gates 2 and 6.

*Logic Network (a)*

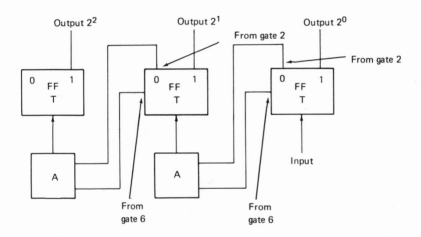

The two AND gates shown in logic network (a) are not additional gates that must be added to the network. Remember that each of the flip-flops is constructed solely from NAND gates, and a NAND gate is an AND gate followed by an inverter. Therefore, the inputs to these flip-flops always feed AND gates. These AND gates can perform the desired ANDing, as shown in logic network (b).

*Timing Chart*

The timing chart for this network is the same as the timing chart for binary counter No. 1.

## Logic Network (b)

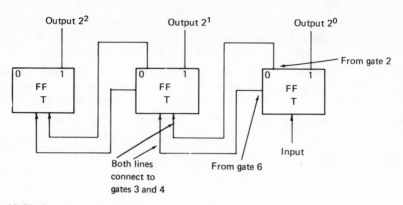

All flip-flops in this network are T type No. 2

## Binary Counter No. 3

A binary counter that has the features of counter No. 2 and in addition may be stepped up by two's or four's.

*Logic Network*

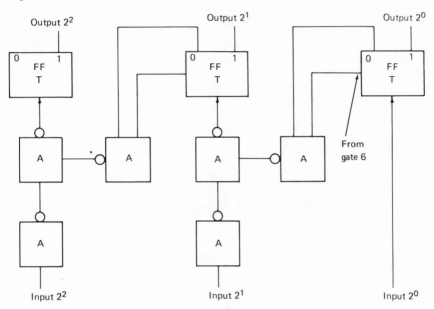

All flip-flops are T type No. 2

*Operational Notes*

1. All inputs are normally down.
2. The raising of any input line will increase the stored count by the power of 2 indicated on the input line.
3. Only one input line should be up at any instant of time.
4. The inputs to all flip-flops are down when all inputs to the counter are down. This condition simplifies resetting and setting operations, as discussed in the section on T flip-flops.

**Counter, Binary Coded Decimal (BCD)**

The following network will count the number of positive pulses appearing on the input line. The count is displayed in BCD code, 0 through 9.

*Logic Network*

*(See Logic Network diagram on page 123.)*

*Timing Chart*

*(See Timing Chart on page 124.)*

*Operational Notes*

1. This network counts the number of positive pulses appearing on the input line, and the output is available in BCD code before the input is lowered.
2. When the counter input line is down, all inputs to flip-flop are down. This feature facilitates setting the network to precount values.
3. An output is available from this network to feed the next decade of a decimal counter. The next decade will be identical to this counter.

*Logic Network*

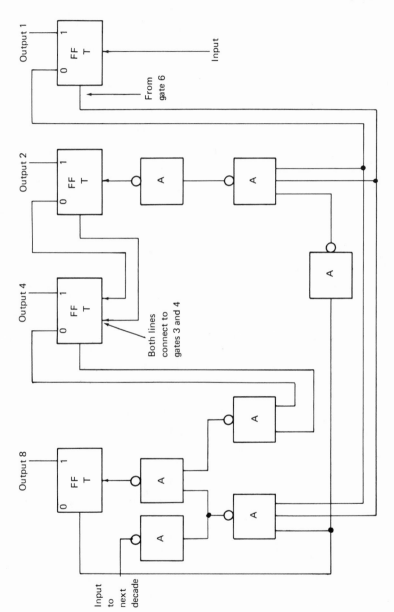

All flip-flops are T type No. 2

123

*Timing Chart*

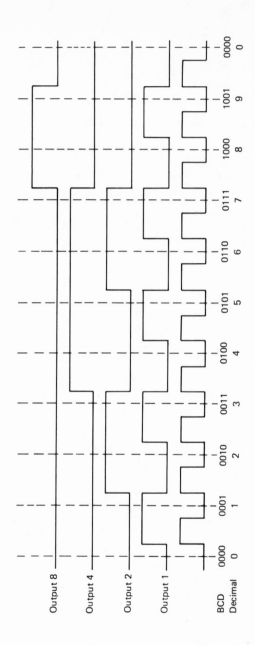

## Counter, Gray Code

The following network will count the number of plus-going pulses appearing on the input line and display the output in Gray Code. Three stages are shown.

*Logic Network*

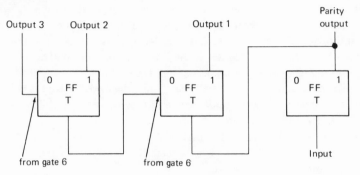

All flip-flops are T type No. 2

*Timing Chart*

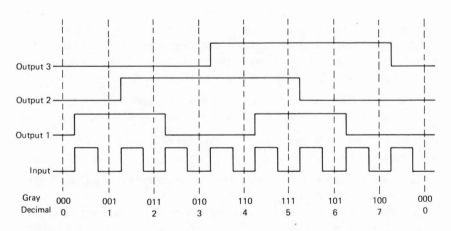

*Operational Notes*

So as not to overly complicate the timing chart, the parity output was not displayed. This output will be up when there are an odd number of other output lines at 1, thus making an even overall parity. This parity bit is a by-product of the network, but it can prove to be very useful for checking purposes.

The counting capability of this network can be extended indefinitely by adding center sections. The first and last flip-flops are connected in a unique manner that does not change with the length of the counter.

**Shift Registers**

*General Information*

A *shift register* is usually defined as a row of flip-flops that has the ability to shift in parallel, left or right, its stored data. This definition is correct, but it tells little about the construction of a shift register. A shift register is actually constructed from two registers and a series of gates that allows the flow of data between the two. One of these registers, called the lower register, is the normal residence of the data. This register is the interface for input and output signals. The second register, called the upper register. is used for temporary storage only. Its function is to contain or store the data while the lower register is being reset and set to its new (displaced) value. The entire operation of a shift register can be explained by a four step process:

1. The upper register is cleared, reset to all 0's.
2. The data from the lower register is gated to the upper register.
3. The lower register is cleared.
4. The data from the upper register is gated to the lower register but displaced one position from its origin.

A simple shift register design can be obtained by directly implementing this four step process with two gated registers and a source of four different clock pulses. This technique is used in the first of the following shift register designs. In this case the design of the actual shift register is simple, but the clocking system is complex. The clocking system can, however, be simplified by using a more complex register design. If an automatic reset type of flip-flop is employed in the design of the registers, then a two clock system is all that is required.

*Clock pulse 1:* Data is gated from lower register to upper register.
*Clock pulse 2:* Data is gated from upper register to lower register (displaced).

This trade of register complexity for clocking simplicity can continue until a one clock system is obtained.

(a) When the clock line is raised, the data from the lower register is gated into the upper register.
(b) When the clock line is lowered, the data from the upper register is gated into the lower register.

It should be pointed out that simplicity of design has nothing to do with the total number of gates required. Shift register No. 7 utilizes a single clock system and yet requires only six gates per position. The network is complex in its feedback connections, but the total number of blocks is minimal.

The following pages do not show completed shift registers but show only one position of the upper and lower registers. It is assumed that all positions are identical and that the user will replicate the network. The author has used the term *shift cell* to mean this one iterative slice of a shift register. In some cases the upper register position of the shift cell is identical to the lower register position of the shift cell and in some cases it is not; but in all cases both are shown.

When working with shift registers the designer often encounters the problem of designing a register that will selectively shift in either direction, or the designer may need to provide a choice of shifts in the same direction. The author calls this type of operation *multimode shifting* and has provided the necessary modifications to each of the following networks to provide for this capability.

## Shift Cell No. 1

A shift cell, basic in design, that is driven by a four line clock system.

### *Logic Network*

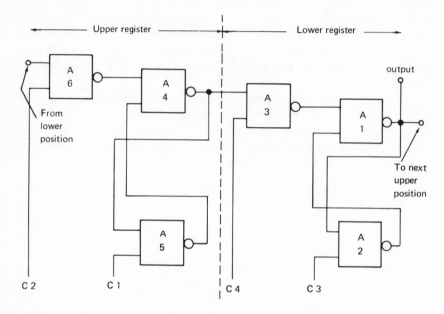

### *Clock System*

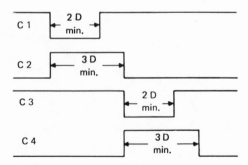

D = delay per gate

This series of pulses can be spread apart timewise as long as the minimum timings are at least maintained. Under no circumstances should C 2 and C 4 be up at the same time.

*Operational Notes*

1. The lowering of C 1 resets the upper register.
2. The raising of C 2 gates the data from the lower register into the upper register.
3. The lowering of C 3 resets the lower register.
4. The raising of C 4 gates the data from the upper register into the lower register.

*Additional Inputs*

Data may be entered directly into the lower register by adding set and reset lines to gates 1 and 2, respectively. These lines should be up when not in use.

*Multimode Shifting*

Gate 6 must be duplicated as shown if shifting in more than one mode is required.

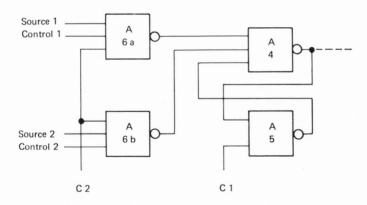

With this arrangement of gating, the upper register will now accept its next value from either of two sources, depending upon which control line is down. The data control lines should be changed only when C 2 is down if minimum timing pulses are being used.

The author elected to put the selective input gating on the upper register, but it could have been put on the lower register. For shifting selectively in many modes, input gating can be added to both registers.

## Shift Cell No. 2

This shift cell is similar to shift cell No. 1 except all required clock pulses are of the same polarity and minimum duration.

*Logic Network*

*(See Logic Network diagram on following page.)*

*Clock System*

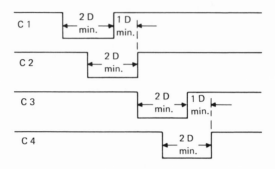

D = delay of one logic gate

This series of pulses may be spread apart timewise as long as the minimum timings are at least maintained. Under no circumstances should C 2 and C 3 be down at the same time, just as C 1 and C 4 should not be down at the same time.

*Operational Notes*

1. The upper register is set to 1 by the lowering of C 1 (output of gate 4 is set to 1).
2. The upper register is selectively reset by the lowering of C 2. The output of gate 4 will now match the input line (line into gate 6).
3. The lower register is set to 1 by the lowering of C 3.
4. The lower register is selectively reset by the lowering of C 4.

*Additional Inputs*

Set and reset lines may be added to gates 1 and 2, respectively, for the purpose of entering data into the lower register. These lines should be up when not being used to set or reset the latch.

*Logic Network*

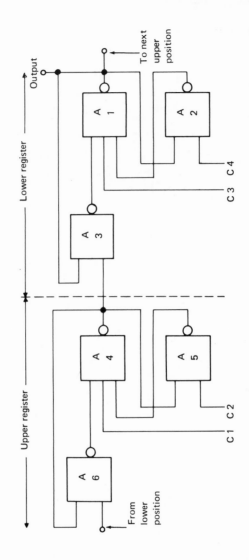

## Multimode Shifting

Gate 6 must be duplicated as shown if shifting in more than one mode is required.

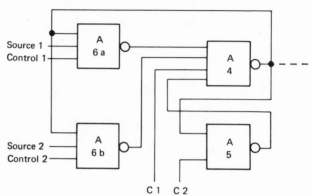

This modification to the shift cell will enable it to accept data from either of two sources, depending upon which control line is up. The value of the control lines should be changed only when C 2 is up if minimum clock pulses are being used. Again, this type of gating could be added to the lower register.

## Shift Cell No. 3

A shift cell designed from a pair of flip-flops of the automatic reset type. Shift registers of this type require a two clock system.

### Logic Network

*(See Logic Network diagram on following page.)*

While eight gates are shown in the network only six are required for each stage, since the inverters, X and Y, are shared among many cells. It should be noted that inverter X connects to upper register gates only, and inverter Y connects to lower register gates only.

### Clock System

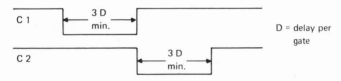

### Operational Notes

This shift cell is of the same design as shift cell No. 1 except it uses a different clocking system. This new system, while simple in design, does introduce the possibility of a race condition. Each of the flip-flops now has its reset signal (inputs to gates 5 and 2) being removed only one gate delay before the set signal (inputs to gates 6 and 3) is removed. This is no problem except that care must be exercised in the powering of lines to gates 5 and 2 so as not to further delay the removal of the reset signals.

Clock lines C 1 and C 2 should never be down at the same time.

### Additional Inputs

Data may be entered directly into the lower register by adding set and reset lines to gates 1 and 2, respectively. These lines should be up when not in use.

### Multimode Shifting

Gate 6 must be duplicated if shifting in more than one mode is required. Refer to shift cell No. 1 for drawing of modifications.

*Logic Network*

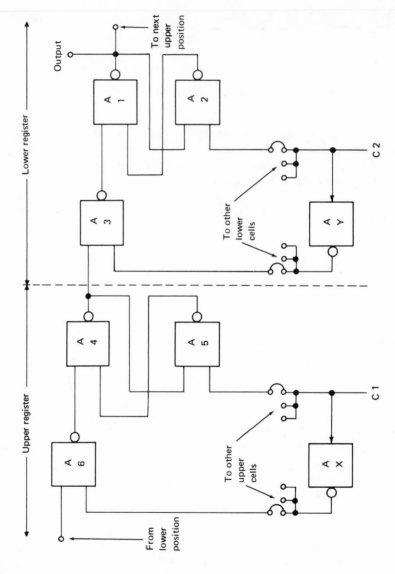

## Shift Cell No. 4

The following network depicts a widely used two clock system shift cell. The simplicity of design accounts for its wide use.

*Logic Network*

*(See Logic Network diagram on following page.)*

*Clock System*

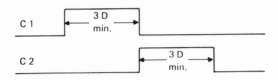

D = delay of one logic gate

Under no circumstances should C 1 and C 2 be up at the same time. The pulses on C 1 and C 2 may be spread apart timewise.

*Operational Notes*

1. When line C 1 is raised, data enters the upper registers.
2. When line C 2 is raised, data enters the lower registers.

*Additional Inputs*

Set and reset lines can be added to gates 1 and 2, respectively, to permit the entering of data into the lower register. These lines should be up when not in use.

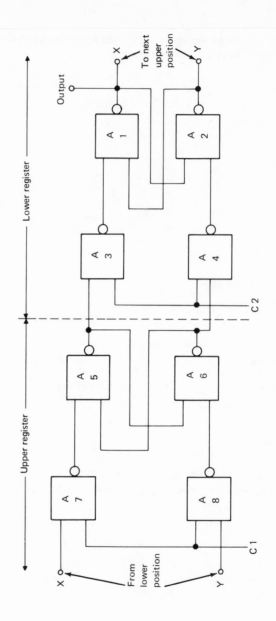

*Logic Network*

## Multimode Shifting

Gates 7 and 8 must be duplicated if shifting in more than one mode is required.

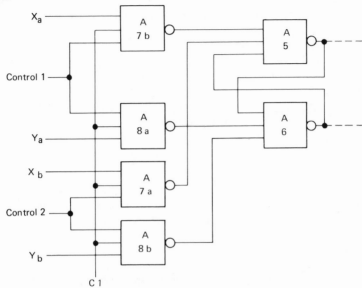

This modification enables the shift cell to accept data from either of two sources, depending on which control line is up. The value of the control line should be changed when C 1 is down if minimum clock pulsing is being used. Selective gating can be added to the upper register.

**Shift Cell No. 5**

A two clock shift cell similar to shift cell No. 4 except only one data line must be connected between cells. This modification reduces the operational speed per shift from six units to seven units, where a unit is the delay per gate.

*Logic Network*

*(See Logic Network diagram on following page.)*

*Clock System*

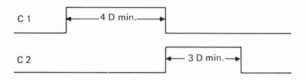

Under no circumstances should C 1 and C 2 be up at the same time. The pulses on C 1 and C 2 may be spread apart timewise.

*Operational Notes*

1. When line C 1 is raised, data enters the upper register.
2. When line C 2 is raised, data enters the lower register.

*Multimode Shifting*

When shifting in more than one mode is required, an additional gate per mode must be added, as shown.

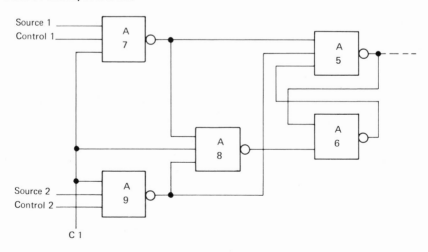

The control line that is up will select the source. The value of the control lines should be changed only when C 1 is down.

*Logic Network*

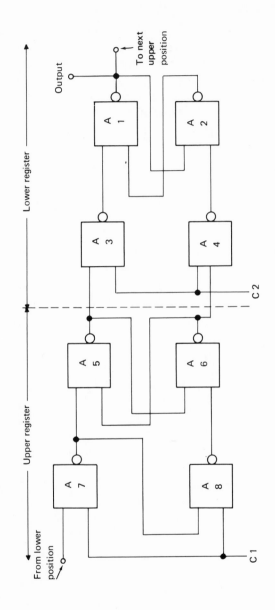

## Shift Cell No. 6

A six gate shift cell requiring only one clock line. This network and the following shift cell were obtained by a formal sequential design procedure. While the procedure is very effective, it produces networks that are difficult to partition into functional units such as upper and lower registers. To obtain a working understanding of networks of this type, one should carefully study the flow chart and the operational notes concerning the network.

*Flow Chart*

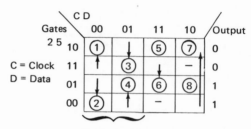

Network resides in either of these two columns when not being stepped

*Logic Network*

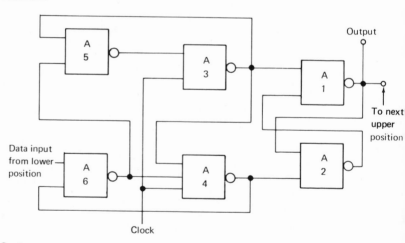

*Clock System*

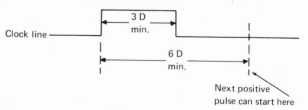

*Operational Notes*

1. When the clock line is down, the output of the network in insensitive to changes on the data line. The network is either locked in the ① ③ states with a 0 output or in the ④ ② states with a 1 output.
2. When the clock line goes up, the network will sample the data line and adjust the output to match. If the data line is 1 when the input is raised, the network will move to stable state ⑥ . If the data line is 0 when the input is raised, the network will move to stable state ⑦ .
3. Once the network reaches stable state ⑥ or ⑦ , a change on the data line will no longer affect the output of the cell.

In summary, the network samples the data line on the rise of the clock line and then becomes insensitive to this input. Remember, this input is in itself an output from another shift cell; and it may start to move as that cell adjusts to its input. Proper operation is based on the fact that the input becomes insensitive before the output changes.

*Additional Inputs*

Set and reset inputs may be added to gates 1 and 2, respectively. These inputs should be up when not in use and should be lowered only when the clock line is down.

*Multimode Shifting*

Gate 6 must be duplicated if shifting in more than one mode is required.

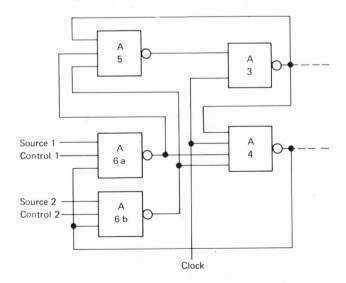

Clock

## Shift Cell No. 7

A six gate shift cell requiring only one clock line. This network can be operated at a slightly higher rep-rate than shift cell No. 6, but it does require two interconnecting data wires between stages.

*Flow Chart*

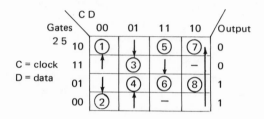

*Logic Network*

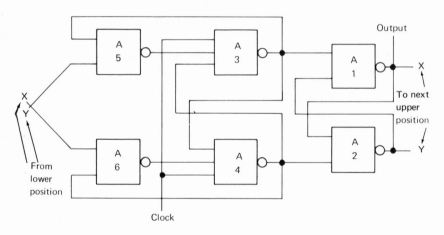

*Clock System*

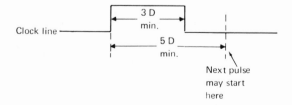

*Operational Notes*

The operational notes for this shift cell are the same as those for shift cell No. 6.

*Additional Inputs*

Set and reset lines may be added to gates 1 and 2, respectively. These inputs should be up when not in use and should be lowered only when the clock line is down.

*Multimode Shifting*

Gates 5 and 6 must be duplicated if shifting in more than one mode is required.

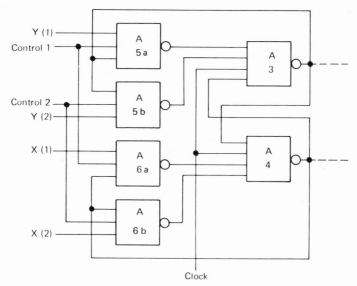

The multimode shift cell will accept data from the source lines that have their corresponding control lines up.

## Ring Circuit or Ring Counter

*General Information*

Ring networks are constructed from an iterative row of bistable stages assembled in "ring" fashion. Each stage connects only to its left and right neighbors and a common clocking line. Usually only one of the stages is set (output at 1) and the remaining stages are reset (outputs of 0). Each time a pulse appears on the clock line the set state progresses one stage at a time around the ring.

Networks of this type were originally designed for counting purposes, and thus the name ring counter. But today these networks have many applications, and many variations of the basic network have been developed. As an example, the ring need not be closed; that is, the last stage need not connect to the first stage. The set state will then proceed from the first stage to the last and disappear. The first stage must then be set to 1 to recycle the network. Some ring networks allow for the setting of any number of stages to 1. At this point, the author prefers to call networks of this type *shift registers*. The networks here labeled ring counters can contain more than one set stage, but it will never be possible to have two adjacent stages set to 1. This results from the design procedure of using a set stage to reset the stage directly to the left. When the clock signal moves the set condition to the next right stage, this new set stage then resets the left (source) stage. This procedure leads to economy of design but rules out use of two adjacent 1's. When two adjacent 1's are required in a ring, refer to the section on shift registers.

## Ring Circuit Type 1

The following ring network requires a two clock system for advancing the ring. The two clock lines are alternately raised and lowered, advancing the ring one step for each complete pulse on either line. This network uses three NAND gates per stage, as do all ring circuits in this section. This network has the fewest interconnections and serves as the basic design for the three following ring circuits.

This network has been designed by interconnecting a series of gated flip-flops. For further information on the basic flip-flop refer to flip-flop, gated No. 1.

*Logic Network*

*(See Logic Network diagram on page 146.)*

*Timing Chart (Four Stages Shown)*

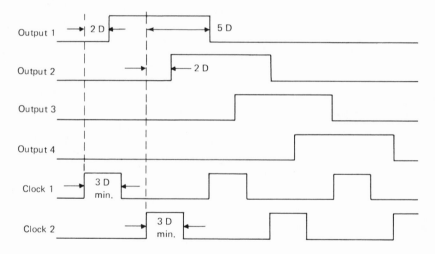

*Logic Network (Two Stages Shown)*

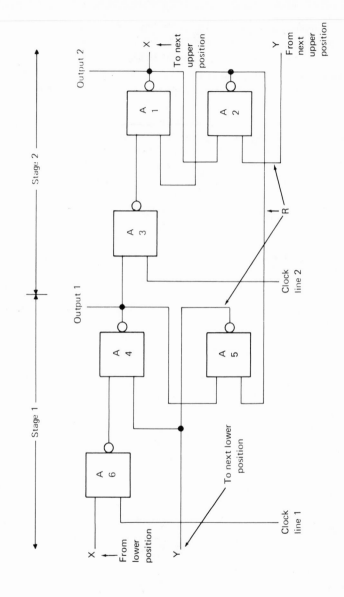

*Operational Notes*

1. When a clock line is raised (either clock 1 or 2), data from every other flip-flop is gated to its right neighbor. Gate 3 performs this gating function when clock line 1 is raised, and gate 6 performs this same function on alternate stages when clock line 2 is raised.
2. When any flip-flop is set to 1, it automatically transmits a reset signal, by way of line R, to its left neighbor. This reset signal resets the flip-flop which was the source of the original set signal. This reset signal is a negative level on line R, and it is not removed until the set flip-flop (the flip-flop from which it is generated) is itself reset. This technique of using a set flip-flop to force a reset on its left neighbor is what makes it impossible to maintain two adjacent 1's in the ring. This technique, however, leads to economy in design.
3. Since the operating principle of this ring is to gate forward and then to receive back a reset signal, it is clear that adjacent output lines will have overlapping up signals. This overlapping of outputs will last for three units of delay regardless of the clock timings.
4. The network is stable with both clock lines down as well as with one clock line up. Both clock lines should never be up at the same time. The nonoverlapping invert of this manual may be used as a source of drive pulses.

*Restrictions*

1. Under no circumstances should clock line 1 and clock line 2 be up at the same time.
2. Network will not maintain 1's in two adjacent stages.
3. This network can be used only when an even number of stages is required. This is a result of the two clock system.

*Additional Inputs*

All positions of the ring can be reset to 0 outputs by adding normally up reset lines to gates 2 and 5. The position that is to be set to a 1 output should have a normally up line connected to gate 4 or gate 1: gate 4 if clock line 1 is up when setting, or gate 1 if C 2 is up.

**Ring Circuit Type 2**

This network shows a modification that can be made to ring circuit 1 to enable it to operate on a two clock system which has overlapping up levels. By adding one wire and one gate input to each stage, less care need be exercised in generating the clock signals. This trade may or may not be advantageous.

*Logic Network*

*(See Logic Network diagram on following page.)*

*Timing Chart*

Only the two clock lines are shown to point out the overlap problem, which has been solved.

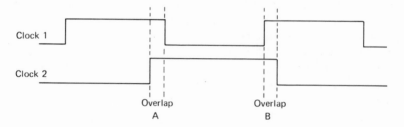

*Operational Notes*

The problem with overlapping clock pulses is that all gates in the ring will be open at the same time. The set stage will advance at an uncontrolled rate, not stopping until the overlap has disappeared. To avoid this problem, line A and line B have been added.

Line A is used to avoid a problem caused by overlap A in the timing chart. Line A goes down when a 1 is being gated into stage 1. The lowering of this line disables gate 3 so that the set condition cannot propagate past stage 1 until clock line 1 is lowered. Line B is used to avoid the same problem caused by overlap B. In this way, the set stage cannot propagate more than one stage at a time.

*Restrictions*

1. Network will not maintain 1's in two adjacent stages.
2. Network can be used only when an even number of stages is required.

*Additional Inputs*

Same as ring circuit type 1.

*Logic Network (Two Stages Shown)*

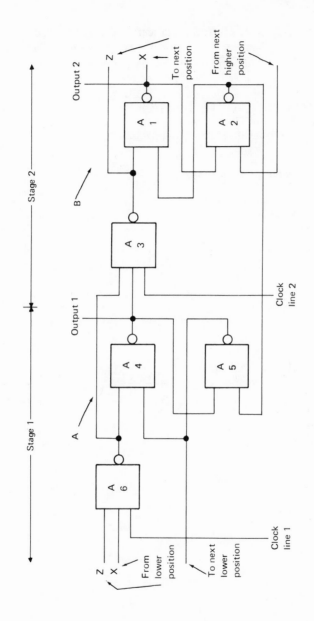

**Ring Circuit Type 3**

This network depicts a further modification that can be made to ring circuit 2 to enable it to operate from one clock line. These modifications reduce the operating speed by adding one extra gate delay per cycle.

*Logic Network*

*(See Logic Network diagram on following page.)*

*Timing Chart (Four Stages Shown)*

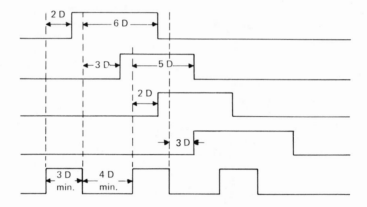

*Logic Network (Two Stages Shown)*

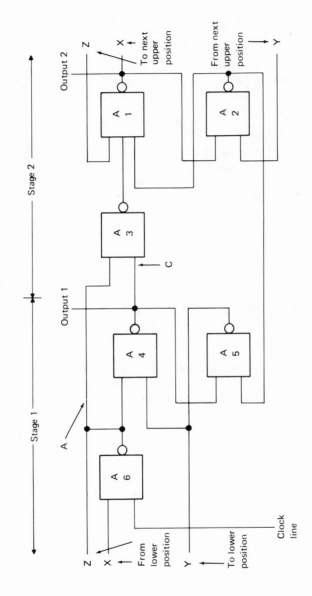

*Operational Notes*

As explained in ring circuit 2, a method has been found which permits the overlapping of the two clock pulses that drive gates 6 and 3. Now, with the ability to tolerate a large variation in the clock timing, it is possible to look for a new source of pulses to replace clock 2. This second clock signal can be obtained from the output of the previous flip-flop, line C in the logic network. Lines A and C combine in gate 3 to:

*first*, inhibit the flow of a set condition to stage 2 while stage 1 is being set by a raised clock line;

*second,* permit the flow of a set condition to stage 2 when stage 1 is set and the clock line goes down.

By connecting clock line 2 to gate 4, a problem has been created in holding down line A for the period of the positive clock pulse. Remember, line A must be held down when the clock line is up to inhibit the uncontrolled forward flow of the set condition. The problem is this:

(a) Stage 1 is set by a positive clock line.
(b) Line Y—the automatic reset line—goes to 0, feeding a reset signal to the source flip-flop.
(c) The source flip-flop is reset, lowering line X.
(d) When line X goes to 0, line A moves up, generating the problem which must be corrected.

The correction consists of connecting line A back to the upper gate of the source flip-flop This forms a new flip-flop—gates 6 and 1—which holds line X up for the duration of the positive clock period. This new flip-flop is reset by the clock line as it normally feeds gate 6.

*Restrictions*

1. Network will not maintain 1's in two adjacent stages,
2. Network can be used only when an even number of stages is required.

*Additional Inputs*

The lowering of a normally up line that is connected to gates 2 and 5 will reset all stages to 0. Only stage 2 can be set when the clock line is down, and this requires the lowering of a normally up line connected to gate 1.

## Ring Circuit Type 4

This network, unlike the previous ring circuits, can be used to construct a ring having an odd number of stages. This feature results from the design which enables ring advancement on the rise of the clock line only.

*(See Logic Network diagram on following page.)*

*Timing Chart (Three Stages Shown)*

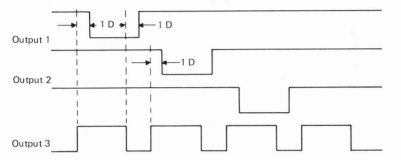

*Operational Notes*

The operation of this network is similar to ring circuit 2. To disable the forward flow of the set condition when the clock line goes down, an additional clock input has been added to gate 3. This input transforms gate 3 into an identical match with gate 6 and so must feed forward an inhibit signal and feed backward a locking signal.

*Additional Outputs*

Gate 3 from each stage may be used as an output in place of gate 1. It is understood that gate 6 is in actuality a gate 3 for stage 1. Using each gate 3 as an output, the following timing chart is obtained.

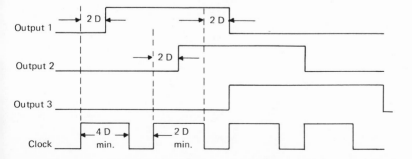

*Logic Network (Two Stages Shown)*

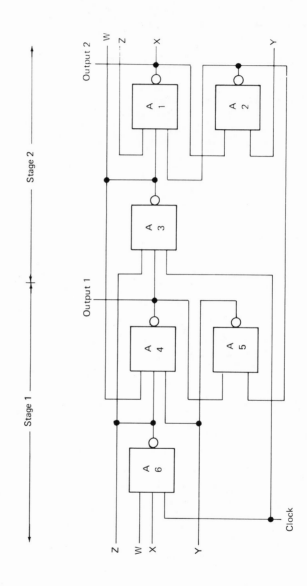

# 4

# AND-OR-INVERT LOGIC

# AND-OR-INVERT

## General Information

The AND-OR-Invert gate is a powerful logic connective yet a difficult one to work with. The difficulty with this connective lies in its flexibility and its many alternate solutions to any particular network design. To further complicate the problem, there is no formal definition of minimality for this connective. When working with NAND logic a simple gate count was an effective criterion of minimality. But when working with AOI connectives one must know more about the packaging of these connectives to establish the relative value of any network. As an example, the AOI gate may be packaged so that three AND gates are connected to the OR-Invert (see Figure 4.1). In this case the saving of one of these AND gates is of no importance, since all three are incorporated into the basic design. On the other hand, the packaging may be such that only "used" ANDs are counted. This occurs when the first AND is packaged with the OR-Invert and additional ANDs are obtained from another package. Packaging of this type is prevalent in industry (see Figure 4. 2). It should be noted that the outputs of package 2 may be connected to the OR inputs of package 1 only. This is a common circuit restriction, as is the rule that no inputs can connect directly to the OR but must first pass through an AND gate. This restriction results in the use of many single input ANDs. The two part package has the following important advantage that should not be overlooked. If the X inputs on package 1 are ignored, then this package becomes a NAND gate; and all NAND networks can be implemented directly with this gate. In a similar manner the AOI gate may be used as an OR-Invert gate. This is accomplished by using only single AND gates, as shown in Figure 4.3. While this OR-Invert design technique is seldom advantageous, on occasion the designer may wish to check over the OR-Invert networks in Section 5 before selecting an implementation.

A single input inverter makes a perfect companion gate for the AOI circuit and is usually supplied with the technology. The networks in this manual assume the availability of this inverter.

## Karnaugh Mapping

Mapping for this connective is a relatively simple procedure requiring the checking of two solutions. Both solutions are obtained from the same map by using the following procedure.

*First:* Draw the usual Karnaugh map from the given truth table.
*Second:* Draw implicant loops around the 0's in the map. The reason for looping the 0's rather than the 1's is to first obtain the inverse function. This

inverse function is then passed through the always present output inverter to generate the true function.

*Third:* Identify the selected loops, and each term then becomes an AND to the AOI. See Figure 4.4. Maps of this type will be shown above combinational networks. The 1's in the map indicate what the function is and the looped 0's indicate how it was obtained.

*Fourth:* The alternate solution is obtained by looping the 1's and then generating an AND-OR network from these loops. Now if the output is double inverted there will be no Boolean effect, and this is precisely what is done. The first inverter is present in the AOI gate so a final single input inverter is added, as shown in Figure 4.5.

While the double inverter solution appears wasteful, it is used frequently in sequential networks. This is due to the fact that a flip-flop always requires two inverters, and one AOI plus one invert is usually considered less expensive than two AOIs.

To repeat, a NAND solution should always be checked when working with AOI gates. NAND solutions are not redrawn in the AOI section, and the user should refer to the AND-Invert section for alternate solutions.

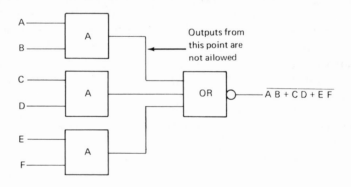

**Fig. 4.1**  AND-OR-Invert gate

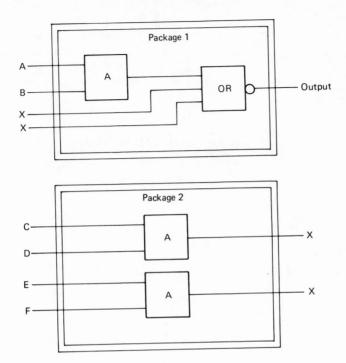

Fig. 4.2  AND-OR-Invert gate partitioned into two packages

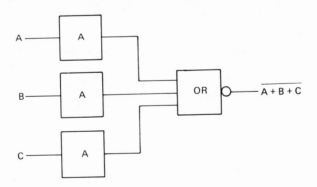

Fig. 4.3  AOI gate used as an OR-Invert gate

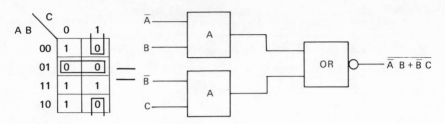

**Fig. 4.4**  Mapping for AOI gates

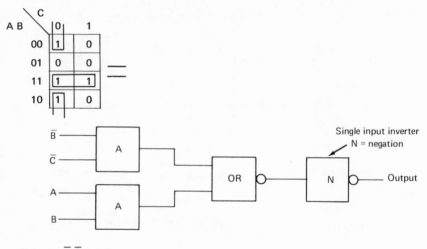

Output = $\overline{B}\ \overline{C} + A\ B$

**Fig. 4.5**  AOI mapping for double inversion

## AOI COMBINATIONAL NETWORKS

### Exclusive-OR No. 1

Inputs are required in true and complemented form for this network.

*Operation*

The output of this network is up when one and only one input is up.

*Logic Network*

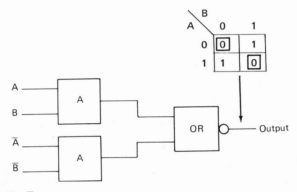

Output = A $\overline{\text{B}}$ + $\overline{\text{A}}$ B

## Exclusive-OR No. 2

Input complements are not required by this network, and the output signal is delayed by two AOI gates.

### *Operation*

The output of this network is up when one and only one input is up.

### *Logic Network*

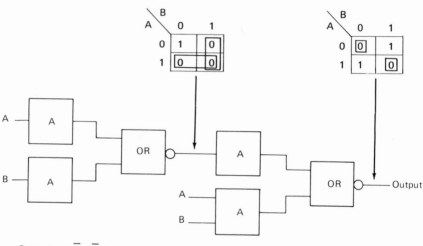

Output = $A \overline{B} + \overline{A} B$

**Exclusive-OR Complement**

Input complements are not required for this network, but the output signal is delayed by two AOI gates. The advantage of this network over the Exclusive-OR network is a saving of one AND gate.

*Operation*

The output of this network is up when both inputs are up or both inputs are down.

*Logic Network*

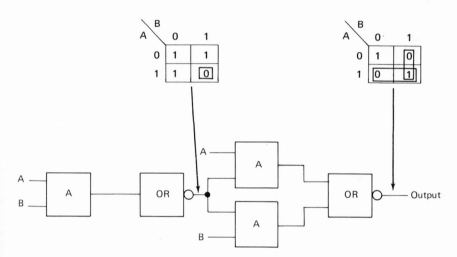

Output = A B + $\overline{A}\ \overline{B}$

*Design Note*

Notice how the AOI on the left is used as a NAND to inhibit both loops in the final map.

## Majority Circuit (Voter Circuit)

To increase the reliability of a switching network, a design procedure called TMR (triple modular redundancy) is used. With this procedure all networks are tripled and the true output is determined by majority rule. The following network will determine the majority of three inputs.

### Operation

The output will be up if any two or all three of the inputs are up. If two or three of the inputs are down, then the output will be down.

### Logic Network

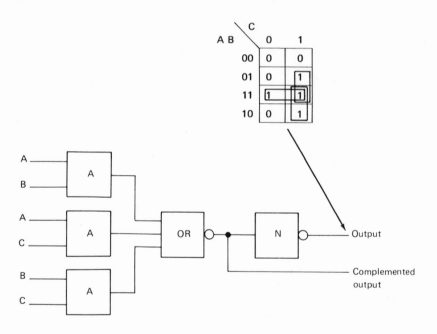

Output = A B + A C + B C

### Design Note

If all inputs to this network are complemented, then the output need not be complemented by the additional inverter.

**Dissent Circuit**

When working within a TMR system, a voter network is used to determine the correct output. But the voter does not indicate the status of the three parallel networks. For maintenance reasons it is desirable to know if one of the three units is in disagreement with the other two, and that is the function of a dissent circuit.

*Operation*

The output of this network will be up if and only if all three inputs agree.

*Logic Network*

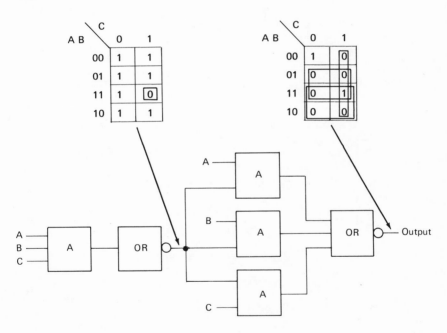

Output = A B C + A̅ B̅ C̅

*Design Note*

Notice how the AOI gate on the left is used as a NAND to inhibit the term A B C from each of the loops of 0's in the final map.

### Odd Circuit—Three Input No. 1

Input complements are not required for this network.
Signals A and B are delayed by four AOI gates.

*Operation*

The output of this network is up when an odd number of inputs is up.

*Logic Network*

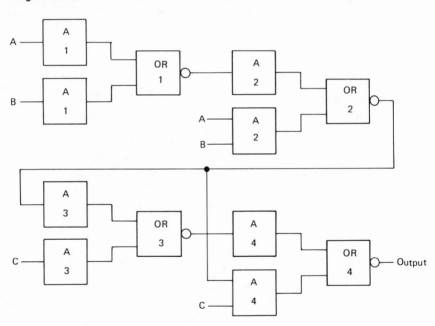

Output $= A\ \overline{B}\ \overline{C} + \overline{A}\ B\ \overline{C} + \overline{A}\ \overline{B}\ C + A\ B\ C$

*Design Note*

This network was designed by connecting two Exclusive-OR networks in series. Gates 1 and 2 comprise one Exclusive-OR and gates 3 and 4 the other.

## Odd Circuit—Three Input No. 2

No input complements are required for this network.
Input signals A and B are delayed by four AOI gates.
This network requires two less AND gates than odd circuit No. 1

*Operation*

The output of this network is up when an odd number of inputs is up.

*Logic Network*

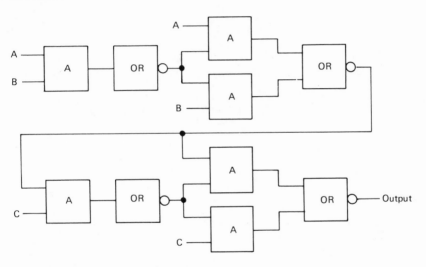

Output = A $\overline{B}$ $\overline{C}$ + $\overline{A}$ B $\overline{C}$ + $\overline{A}$ $\overline{B}$ C + A B C

*Design Note*

This network was designed by connecting two Exclusive-OR complement networks in series. It is interesting to note that the same function is obtained by connecting an even number of Exclusive-ORs or Exclusive-OR complement networks in series.

### Even Circuit—Four Input

The following four input even circuit requires four AOI gates. The output signal is delayed by four gates just as it is delayed in an odd circuit.

The output of this network is up when no inputs, two inputs, or four inputs are up.

*Karnaugh Map*

| A B \ C D | 00 | 01 | 11 | 10 |
|-----------|----|----|----|----|
| 00 | 1 | 0 | 1 | 0 |
| 01 | 0 | 1 | 0 | 1 |
| 11 | 1 | 0 | 1 | 0 |
| 10 | 0 | 1 | 0 | 1 |

*Logic Network*

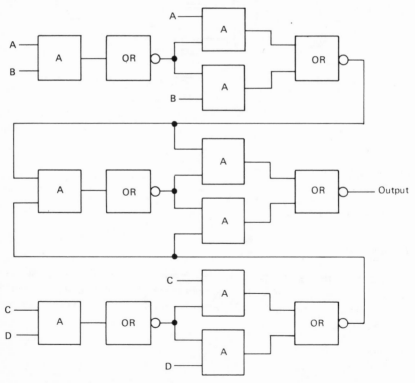

Output = $A B \bar{C} \bar{D} + A \bar{B} \bar{C} D + \bar{A} \bar{B} C D + \bar{A} B C \bar{D} + \bar{A} B \bar{C} D + A \bar{B} C \bar{D} + A B C D + \bar{A} \bar{B} \bar{C} \bar{D}$

## Odd Circuit–Four Input

A four input odd circuit with a gate count of six AOIs. All signals are delayed by four AOI gates and input complements are not required. This network was designed by cascading a series of Exclusive–OR networks.

*Logic Network*

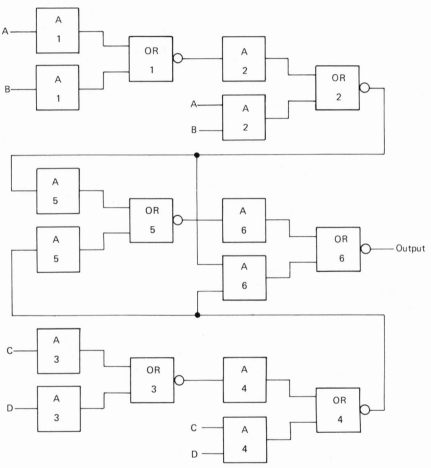

Output = $A B C \overline{D} + A B \overline{C} D + A \overline{B} C D + \overline{A} B C D + A \overline{B} \overline{C} \overline{D} + \overline{A} B \overline{C} \overline{D} + \overline{A} \overline{B} C \overline{D} +$
$\overline{A} \overline{B} \overline{C} D$

**Adder—Half**

A two gate solution not requiring input complements.
Both the sum and carry outputs are inverted.
Inverted carry output signal is delayed by only one AOI gate.

*Logic Network*

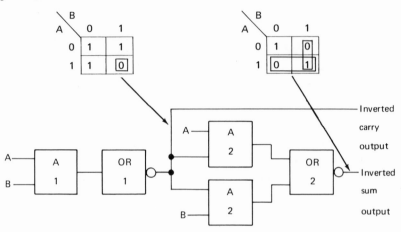

Inverted carry output $= \overline{A} + \overline{B}$

Inverted sum output $= A B + \overline{A}\ \overline{B}$

*Design Note*

Gate 1 is used to generate the term A B. This term aside from being used as an output is also used to inhibit the 1 from the zero implicant loops of gate 2.

### Adder—Full No. 1 (or Even Circuit—Three Input)

This network generates the complement of the sum and carry outputs.

Network requires only two AOI gates, one with four ANDs and the other with three ANDs.

The inverted carry output signal is delayed by only one AOI gate, but it requires inversion before it can be connected as an input to the next higher full adder.

The inverted sum output of this network has the same Boolean function as a three input even circuit and may be used as such. The inverted carry output is ignored in an application of this type.

*Logic Network*

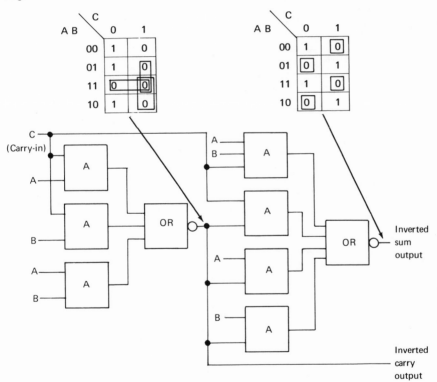

$$\overline{Sum} = \overline{A}\,\overline{B}\,\overline{C} + \overline{A}\,B\,C + A\,B\,\overline{C} + A\,\overline{B}\,C$$

$$\overline{Carry} = \overline{A}\,\overline{B} + \overline{A}\,\overline{C} + \overline{B}\,\overline{C}$$

## Adder—Full No. 2

This adder requires six AOI gates in place of two for the previous adder. However, the sum and carry outputs are in true form, and the carry output can be connected directly to the carry-in of the next adder.

The carry output signal is delayed by two AOI gates before appearing as a carry output signal.

This network was designed by a trial and error procedure.

*Logic Network*

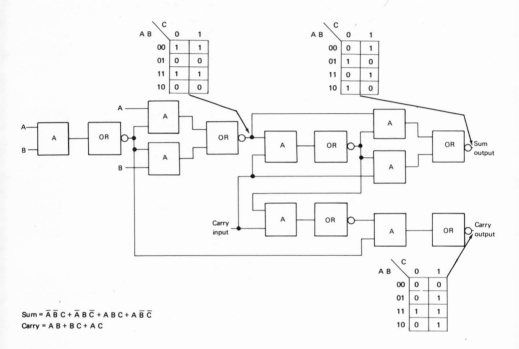

$$\text{Sum} = \overline{A}\,\overline{B}\,C + \overline{A}\,B\,\overline{C} + A\,B\,C + A\,\overline{B}\,\overline{C}$$
$$\text{Carry} = A\,B + B\,C + A\,C$$

**Adder—Connector**

This network will function as a full adder; and with the additional controls, it can generate the Exclusive-OR, AND, or OR of the input variables A and B.

All outputs are inverted including the connective functions.

*Logic Network*

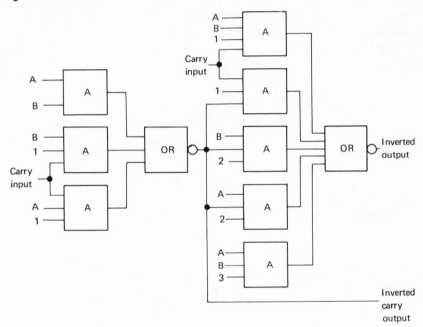

*Control*

For full adder operation
Line 1 up, line 2 up, line 3 down
For inverted Exclusive-OR operation
Line 1 down, line 2 up, line 3 down
For inverted AND operation
Line 1 down, line 2 down, line 3 up
For inverted OR operation
Line 1 down, line 2 up, line 3 up.

## AOI SEQUENTIAL NETWORKS

### Flip-Flop Latch (Flip-Flop, Set/Reset)

The following network is the basic storage element in the AOI logic family. When the set line is down and the reset line is up, the network has two stable states. Unlike the NAND flip-flop, both outputs of this network are always complements of each other when the network is stable. However, the B output is always the first to respond to an input change.

*Flow Chart*

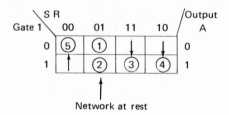

Network at rest

*Logic Network*

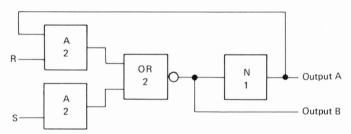

*Operational Notes*

Reset line is normally up and the set line is normally down.

Raising the set line will set the flip-flop; output A will go to 1 and output B will go to 0. Lowering of the set input will have no effect on the outputs.

Lowering the reset input will cause output A to go to 0 and output B to go to 1. Raising the reset input will have no effect on the output.

*Additional Inputs*

Additional set lines may be connected to the network by adding more single input ANDs to the AOI gate. When this is done, the raising of any set line will result in setting the flip-flop (output A = 1, output B = 0).

### Inverter, Nonoverlapping

This network generates a true and an inverted signal which are nonoverlapping in the up level. The timing chart shows that the two outputs of the network are never up at the same time.

This network is particularly useful for driving ring and shift circuits.

*Timing Chart*

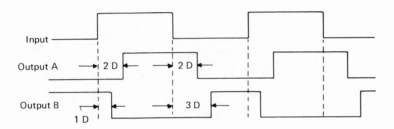

*Flow Chart*

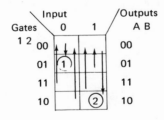

*Logic Network*

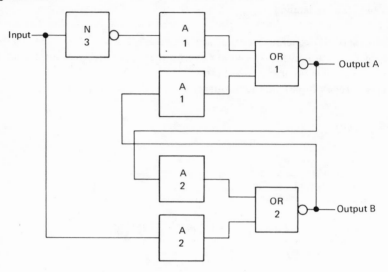

*Operational Notes (From Flow Chart)*

The rise and fall of the input signal moves the operating point back and forth between stable states ① and ② . By following the arrows on the flow chart, it can be observed that the operating point never moves through row 11. This avoidance of row 11 prohibits the two outputs from being up at the same time.

### Flip-Flop, Gated No. 1

A flip-flop with one or more set inputs where each input is under control of a separate gate line.

*Flow Chart*

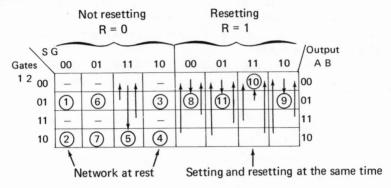

Network at rest        Setting and resetting at the same time

*Logic Network*

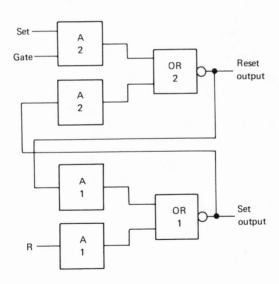

*Operational Notes*

The gate line and the reset line are normally down.

The network is independent of all changes on the set line as long as the gate line is down.

Outputs A and B are complements of each other except when attempting to set and reset the network simultaneously. When this input condition occurs, the network will move to stable state ⑩ where both outputs are down.

The gate line and the set line must both be up for the delay time of two AOIs to insure setting of the latch.

For high speed operation the gate and the reset signals are usually overlapped, as shown below.

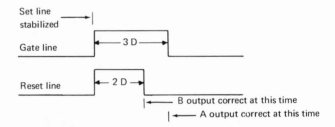

*Additional Inputs*

Additional gated set lines may be incorporated into the design by duplicating the upper AND gate. Only one gate line should be raised at a time, and the network will be independent of all changes on set lines when associated gate lines are down.

**Flip-Flop, Gated No. 2**

A basic flip-flop with one or more gated set lines.

*Flow Chart*

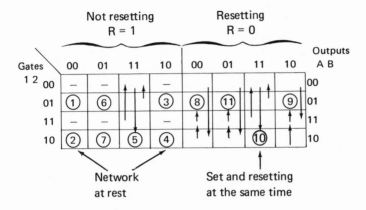

*Logic Network*

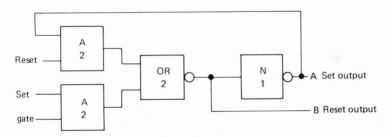

## Operational Notes

The gate line is normally down and the reset line is normally up.

When the network is stable, the outputs are always complements of each other.

When the gate line is down, the network is insensitive to changes on the set line.

Raising the gate line when the set line is up will result in setting the flip-flop.

## Additional Inputs

Additional gated set lines may be incorporated into the network by adding more ANDs to the AOI gate.

## Design Note

The flow chart for this network is identical to that of gated flip-flop No. 1 except for the row location of stable state ⑩ .

**Flip-Flop, Automatic Reset Type No. 1**

Gated flip-flops of the following type are useful in high speed switching networks, for they do not require resetting prior to input gating. When the input gate line is raised, the flip-flop is conditioned to the value of the data line.

*Sample Timing Chart*

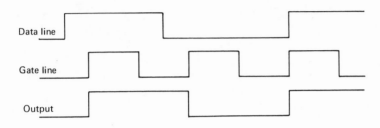

*Flow Chart*

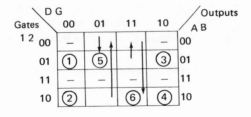

*Logic Network*

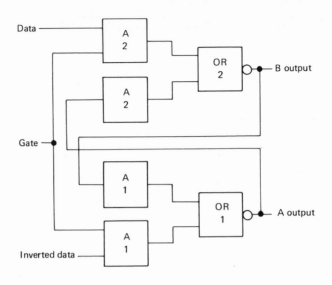

*Operational Notes*

This network requires the availability of the data and its complement. The two output lines are complements of each other except when the outputs are changing, in which case they are temporarily both 0.

When the gate line is down, the network is insensitive to changes on the data line.

When the gate line is raised, the flip-flop will follow the value of the data line.

### Flip-Flop, Automatic Reset Type No. 2

This flip-flop is similar to the previous flip-flop except it has the added advantage of not requiring an inverted data line. The network does, however, require one additional AOI gate over that of the previous network.

*Flow Chart*

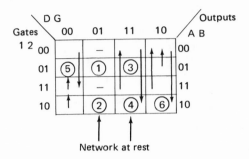

*Logic Network*

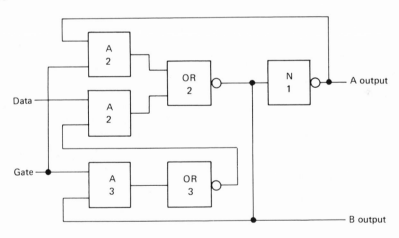

*Operational Notes*

The gate line is normally up and the two output lines are complements of each other.

When the gate line is lowered, the flip-flop will be conditioned to the value of the data line.

As long as the gate line is lowered, the network will follow the value of the data line.

When the gate line is up, the network is insensitive to the data line.

## Flip-Flop, Automatic Reset Type No. 3

This three gate flip-flop is useful when gating data from a number of sources, but it contains a hazard which may be serious. This hazard does not appear on the merged flow chart, but it is discussed in connection with a diagram.

*Flow Chart*

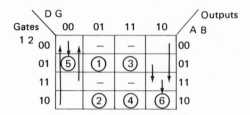

*Logic Network*

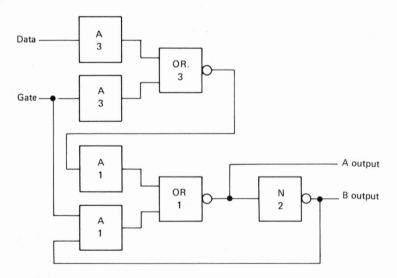

*Network Operation*

Gate line is normally up and both outputs are complements of each other.

When the gate line is lowered, the flip-flop will be conditioned to the value of the data line.

As long as the gate line is down, the network will follow the value of the data line.

When the gate line is up, the network is insensitive to the data line.

*Additional Input*

This network can be modified to accept data from more than one source by adding to gate three, as shown.

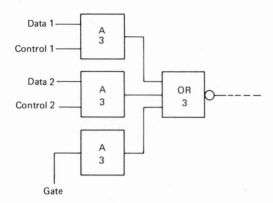

With this modification, the network will accept data from the AND gate whose control line is up when the gate line goes down.

*Hazard*

The hazard in this network can best be demonstrated in the following partial diagram.

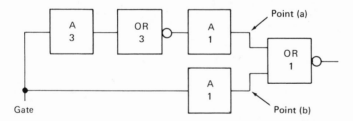

Assume the gate line and point (b) are down and point (a) is up. When the gate line is raised, the network will fail if point (b) does not rise before point (a) falls. This may not appear to be a serious hazard, but it is a hazard that must be watched. The problem is actually in the rise time of the gate line. Gate lines are usually heavily loaded and rise times are often poor. When this condition is severe, the voltage gain in AOI No. 3 may be sufficient to lower point (a) before point (b) reaches the 1 level. In summary, this network cannot tolerate severe degradation of the rise time on the gate line.

### Flip-Flop, Automatic Reset Type No. 4

This flip-flop is constructed from only two gates and does not require an inverted data input. However, it does require two gate lines, one being the complement of the other. The timing between transitions on the two gate lines is not critical.

*Flow Chart*

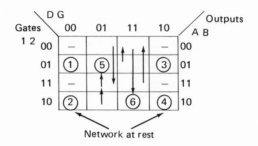

Network at rest

*Logic Network*

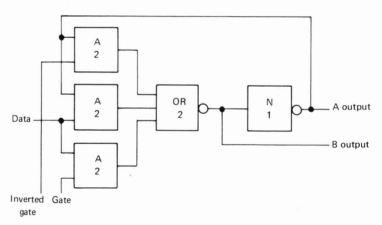

*Operational Notes*

The gate line is normally down and the inverted gate line is normally up.

When the gate line is raised and the inverted gate line is lowered, the network will follow the value of the data line.

When the gate line is down and the inverted gate line is up, the network is insensitive to changes on the data line.

The timing between the gate line and the inverted gate line is not critical, but the data line must remain stable throughout the period of time when the gate line is going down and the inverted gate line is coming up.

## Flip-Flop, Three State

The following network has three stable states when all inputs are down. One and only one of the outputs will be down for each of these states. The network can be moved directly from any state to any other state by raising the appropriate input.

*Partial Flow Chart*

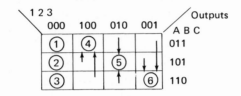

*Logic Network*

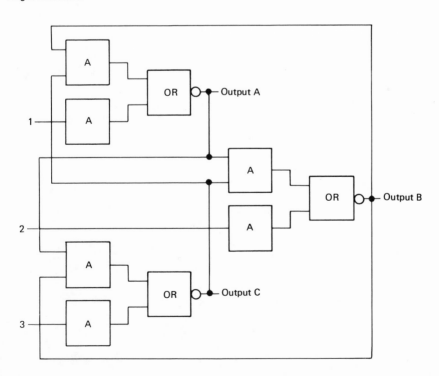

*Operational Notes*

1. All inputs are normally down.
2. One and only one output is down when all inputs are down.
3. The lowered output can be moved by raising an appropriate input.
4. If two inputs are raised, then both related outputs will be held at the lower level. If three inputs are raised, all outputs will be held at the lower level.
5. When more than one input is raised, the final state of the network is determined by the input that is lowered last.

*Design Note*

A four state flip-flop can be obtained by extending the design principle of the network. Use four AOI gates and connect the output of each to all other AOIs.

## Flip-Flop, T Type No. 1

The following T type flip-flop is constructed from four AOI gates plus one single input inverter. Before selecting this flip-flop or the following flip-flop however, it is advisable to check T type flip-flops in the AND-Invert section of this manual.

*Timing Chart*

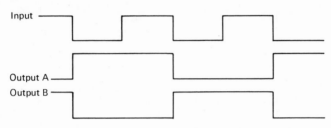

*Logic Network*

*(See Logic Network diagram on following page.)*

*Design Note*

This network was designed by connecting together two automatic reset type flip-flops. The output of the first connects directly to the second, while the second has its output crossed and connected to the first.

*Logic Network*

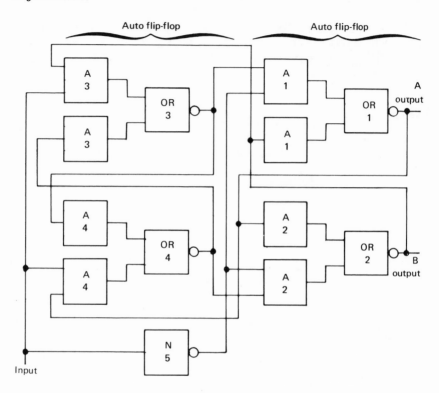

*Operational Notes*

When the input line is down, data from the left flip-flop is gated into the flip-flop on the right. When the input line is up, the data from the right flip-flop is cross gated into the left flip-flop. Cross gating produces the same results as gating the inverted data. In this way the output will alternate between 1 and 0 each time the input is lowered.

This T flip-flop can be modified to change state on the rise of the input pulse by relocating the single input inverter to the drive line of the left auto flip-flop.

### Flip-Flop, T Type No. 2

This network is similar to T flip-flop No. 1 in that it has been constructed from two automatic reset type flip-flops. The automatic reset flip-flops used in this case are different from those used in T flip-flop No. 1, but the input-output timing chart is the same.

*Timing Chart*

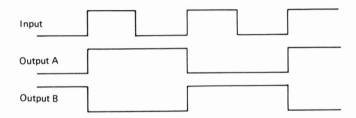

*Logic Network*

*(See Logic Network diagram on following page.)*

*Operational Notes*

When the input line is down, the inverted data from the right-hand auto flip-flop is gated into the left auto flip-flop. When the input line is up, the data from the left auto flip-flop is gated into the right auto flip-flop. In this way the output will change alternately each time the input line is raised.

This T flip-flop can be modified to change on the fall of the input pulse by connecting the X line on the input to the $\overline{X}$ inputs on the network and the $\overline{X}$ input line to the X inputs of the network.

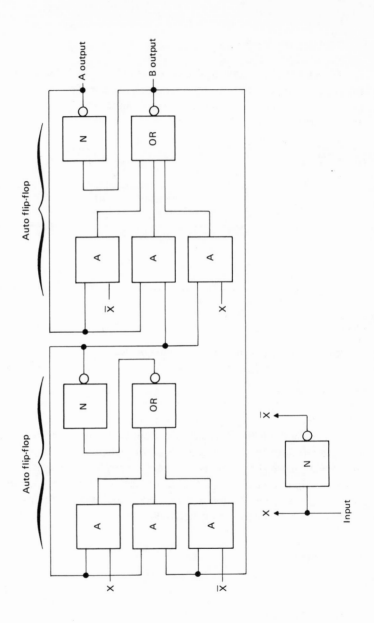

*Logic Network*

## Flip-Flop, Set Dominant

This network operates as a conventional set/reset flip-flop, with the added advantage that it will always set if both set and reset signals are applied at the same time. This set dominant characteristic holds regardless of which input signal arrived first, or is removed first, as long as both signals were present at the same time.

*Flow Chart*

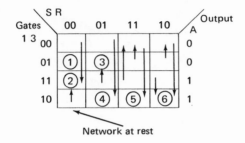

*Logic Network*

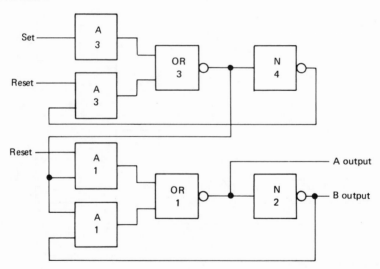

*Operational Notes*

Both inputs are normally down and the inputs are complements of each other.

To set the flip-flop the set input line is raised, and to reset the flip-flop the reset input line is raised.

If both inputs are up at the same time, the flip-flop will be set regardless of which input is lowered first.

## Flip-Flop, Post Indicating

This network differs from a basic flip-flop in that its outputs do not change when setting or resetting is taking place. The outputs change only when the set or reset signal is being removed and thus the name *post indicating flip-flop.*

*Flow Chart*

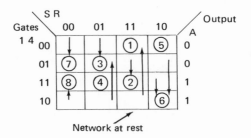

Network at rest

*Logic Network*

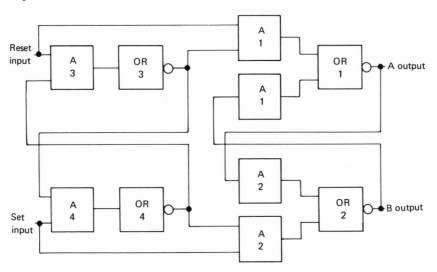

*Operational Notes*

The set and reset inputs are normally up.

Lowering the set line or the reset line will set or reset the flip-flop, respectively, but the output will not change until the lowered input is returned to its normally up position.

## Gated Oscillator No. 1

This network will gate an oscillator signal without shortening or lengthening any of the oscillator pulses. This statement is true no matter when the gate line is raised or lowered, and no fractional pulses will appear on the output.

*Input-Output Timing Chart*

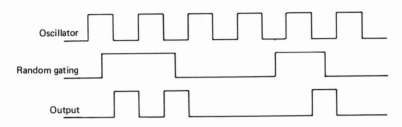

*Flow Chart*

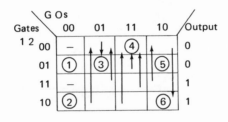

*Logic Network*

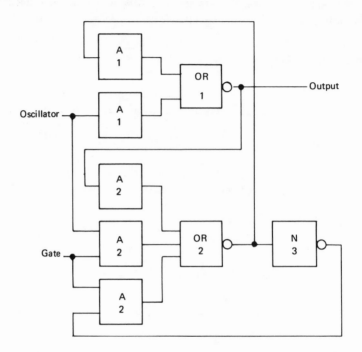

*Operational Notes*

From the above flow chart

1. It is assumed that the oscillator signal is continuously applied to the network. The network will, therefore, move between stable states ① and ③ when the gate line is down. The output will remain 0 as the network moves between these two states.

2. For the output to go up, the operating point must reach state ② or ⑥ ; but the only path to these states is through state ④ to column 10. To get to state ④ the gate line must be up at the same time as the oscillator line is up: column 11. Once in state ④, the network will move in unison with the oscillator input, going first to state ⑥ , back to state ④ , and so on.

3. State ② is used to complete an output pulse when the gate line is lowered while the operating point is at ⑥ .

## Gated Oscillator No. 2

When the gate line is raised, this network will gate one and only one oscillator pulse. The output pulse will always be a full width pulse regardless of the timing of the gate line.

*Input-Output Timing Chart*

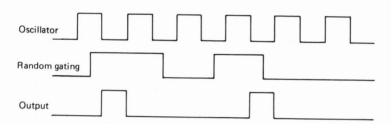

*Partial Flow Chart*

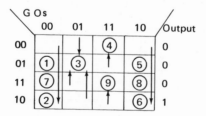

*Design Note*

This network can selectively perform the function of gated oscillators No. 1 and No. 2 by adding a control line to the lower AND in AOI No. 4. By lowering this control line, the effect of the newly added flip-flop can be canceled.

*Logic Network*

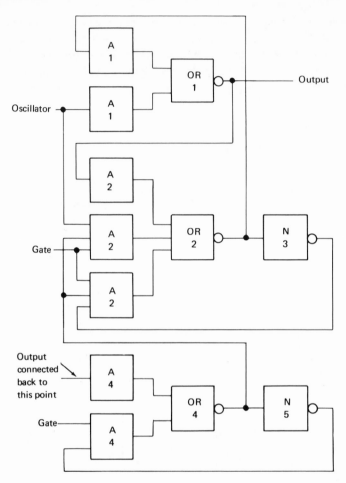

*Operational Notes*

This network is basically the same as gated oscillator No. 1 except that an additional flip-flop has been connected between the output and the AND gates of the input. The output from AOI gate No. 4 of this flip-flop is normally up so it does not block the gate input signal. But when the output line of the network goes up, it will set this flip-flop and gate No. 4 will go down. This will block the gate signal to the network, and no further output pulses will appear until the flip-flop is reset. Resetting takes place, through the AND gate on AOI No. 4, when the gate line is lowered. Thus the gate line must be lowered and raised each time an output pulse is desired.

## One Shot (Single Shot Circuit)

When the input line to this network is lowered, the output line will supply a plus-going pulse of constant time duration. The input line can be returned to its normally up level at any time without affecting the output pulse width. The duration of the output is determined solely by the sum of the delay through one AOI gate and the delay element. The delay element may be a delay line or an RC network.

*Logic Network*

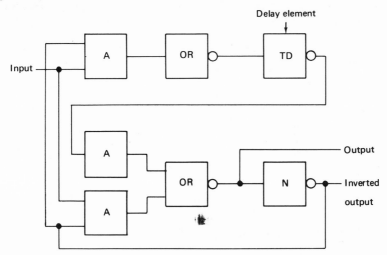

*Timing Chart*

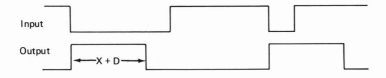

X = Delay imposed by delay element

*Design Note*

After generating an output pulse, the network cannot be recycled until the delay path has cleared. This means that output pulses cannot be any closer together than X + D units.

If the single input inverter and its associated feedback loops are removed, the network becomes a pulse shortener. The output pulse width under this design change will be the smaller of the following two conditions: the input pulse width and X + D delay units.

**Binary Counter**

The following network will count the number of positive (or negative) pulses appearing on the input line. The count is continually displayed in binary, one output line per T type flip-flop. The network can be constructed from T type flip-flop No. 1 or No. 2. The rightmost T flip-flop should be internally connected to change state on the rise of the input pulse if counting of positive pulses is required. Likewise, internally connecting the first flip-flop to change state on the fall of the input line will result in a counting of the negative pulses. All higher order T flip-flops should be internally connected to change state on the rise of their input lines regardless which type of pulses the network is counting.

*Logic Network*

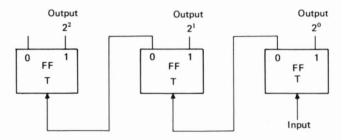

*Timing Chart (for counting of positive pulses)*

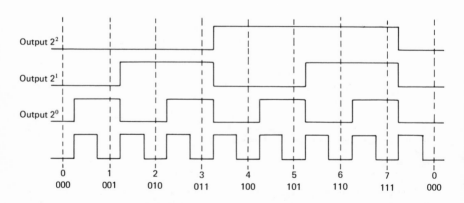

This timing chart points out the value of the output lines when the input line is down, but it should be observed that the output lines actually become stable at these values while the input line is still up.

*Design Notes*

This network can introduce problems to a designer as a result of the method used to interconnect the flip-flops. When the network input line is down, the interconnecting lines may be up or down, depending upon the counted value. This characteristic introduces problems when it comes to setting the counter to predetermined values. Setting or resetting any T flip-flop is always simpler if the input condition is known.

The user is advised to review the counters presented in the AND-Invert section of this manual. It should be remembered that all AND-Invert networks shown earlier can be implemented with AOI gates and this includes T type flip-flops.

**Bidirectional Counter**

This counter network has a controlled directional count. It can count in ascending or descending sequence. The direction of counting can be changed without disturbing the count, and this is an unusual feature.

A single stage of the counter is shown first; then an entire counter is displayed.

*Bidirectional Counter Stage*

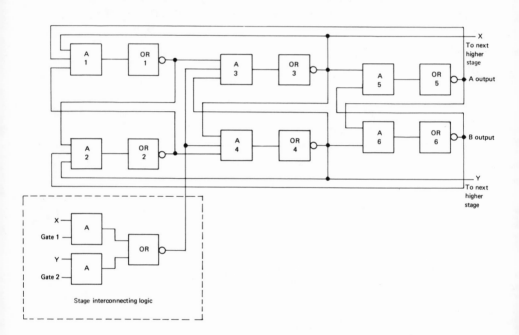

This network is actually a T flip-flop similar to AND-Invert T flip-flop No. 1. It has special internal connections on gates 1 and 2 so as to produce outputs X and Y. These outputs can be thought of as a carry signal (Y) and a borrow signal (X). Both lines are normally up, which is the reason that one gate control line should be raised before the other is lowered when reversing direction of counting.

*Counter Network*

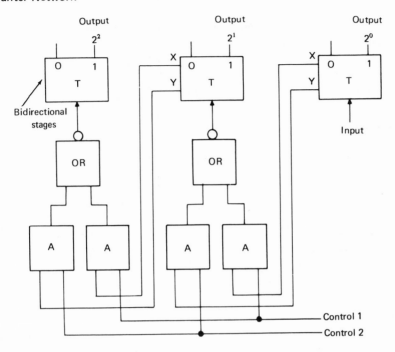

*Operational Notes*

The basic T flip-flop in this network changes state on the rise of the input pulse.

The stage interconnecting lines are normally up. The X output line of a flip-flop will go down when the A output of that flip-flop is moving to 1. It will stay down only as long as the input to that flip-flop is up. The Y line will go down when the A output is 0 and the input to that flip-flop is up.

When control line 1 is up, the X outputs are effective and the counter counts down. When control line 2 is up, the Y outputs are effective and the network counts up.

When changing the control lines, both should not be down at the same time. Raise one before lowering the other.

## Shift Cell No. 1

For general information on shift cells refer to the AND-Invert section of this manual.

The following shift cell is constructed from two identical flip-flops of the automatic reset type.

This network requires a two line clock system.

### *Logic Network*

*(See Logic Network diagram on following page.)*

### *Clock System*

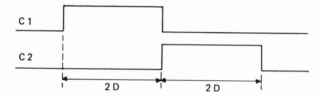

D = Delay through one AOI gate

It is important that both lines should *not* be up at the same time.

### *Operational Notes*

When C 1 is up, the data from the lower register is entered into the upper register.

When C 2 is up, the data from the upper register is entered into the lower register.

Usual practice suggests that data be stored in the lower register when shifting is not taking place. It is, therefore, this register which will require additional set and reset lines for the entry of data to be shifted. This set/reset feature can be incorporated into the design by adding a single input AND to AOI No. 1 and AOI No. 2. The additional input to AOI No. 2 will set the flip-flop when raised, and the input to AOI No. 1 will reset the flip-flop when raised. Both input lines should be down when shifting is taking place.

*Logic Network*

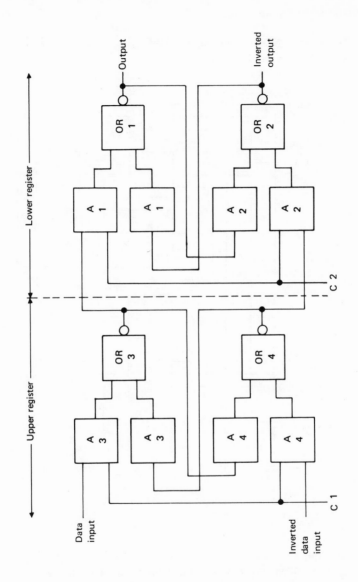

## Multimode Operation

Additional ANDs can be added to AOI No. 3 and No. 4 if shifting in more than one mode is required.

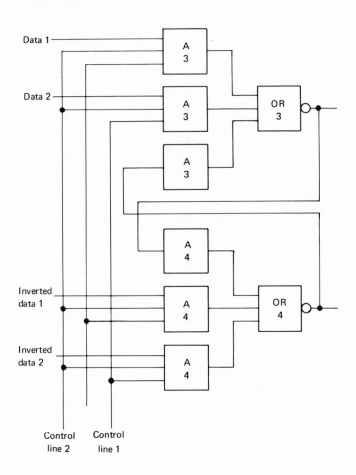

With this modification the shift cell will now accept data from either of two other shift cell positions, depending on which control line is up.

## Shift Cell No. 2

The following shift cell is constructed from a post indicating flip-flop with gated input. It can be partitioned into an upper and a lower flip-flop, but for most applications it is advisable to view the entire network as a gated post indicating flip-flop.

This network requires only one clock line which is normally down when shifting is not taking place.

*(See Logic Network diagram on following page.)*

### *Operational Notes*

The clock line is normally down.

When the clock line is raised, data is gated into the post indicating flip-flop. The outputs of a post indicating flip-flop do not change until the input signals are removed so the output line and the inverted output line do not change state at this time. However, when the clock line is lowered, shutting off the input signals, the output lines move to the value stored when the clock line was up.

Set and reset lines may be added to gates 3 and 4. Normally, these lines should be up. Setting and resetting should take place only when the clock line is down.

With the following modifications the network will accept data from either of two sources.

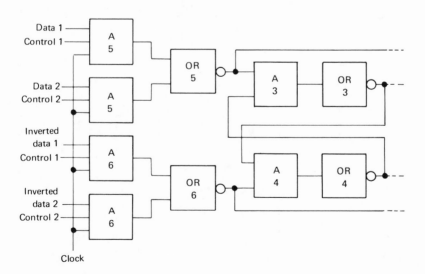

The network will accept data from the set of AND gates whose control line is up.

*Logic Network*

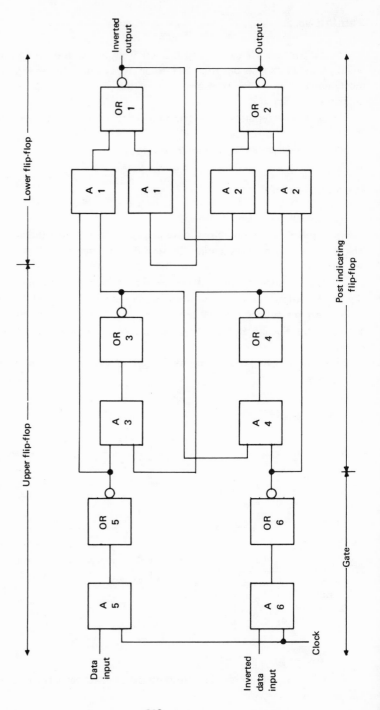

**Shift Cell No. 3**

This shift cell is very similar to shift cell No. 2 except that an inverted data input line is not required. This feature is advantageous in applications where additional gating is required between cells.

This shift cell, like cell No. 2, requires only one clock line, and this line is normally down when shifting is not taking place.

*Logic Network*
*(See Logic Network diagram on following page.)*

*Operational Notes*

The clock line is normally down. When the clock line is raised, data is gated into the post indicating flip-flop. The output line and the inverted output line do not change value until the clock line is once again lowered.

Set and reset lines may be added to gates 3 and 4, as explained in shift cell No. 2.

*Multimode Operation*

When shifting in more than one mode is required, an additional AOI gate must be added per mode. The following diagram shows the necessary modifications for multimode shifting.

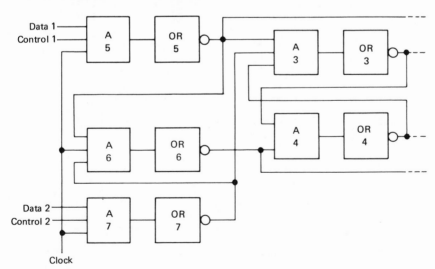

When the clock line is raised, the shift cell will now accept data from the gate whose control line is up.

*Logic Network*

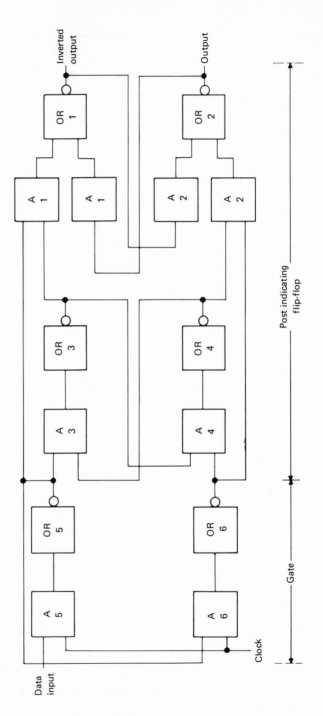

210

## Ring Counter No. 1

For general information on ring counters refer to the AND-Invert section of this manual.

The following ring network requires a two clock system. The two clock lines are alternately raised and lowered, advancing the ring one step for each complete pulse on either line.

This network has been designed by interconnecting a series of gated flip-flops. Each stage is actually one gated flip-flop and the clock system passes the set condition from one flip-flop to the next.

*Logic Network (Two Stages Shown)*

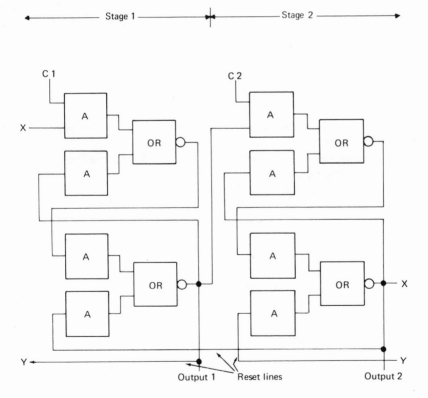

*Timing Chart (Four Stages Shown)*

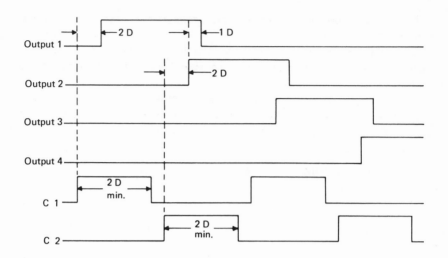

*Operational Notes*

Under no circumstances should C 1 and C 2 be up at the same time. A nonoverlapping inverter should be used as the drive source.

Each time a clock line is raised the set condition is advanced one stage. The setting of this stage generates a signal that is used to reset the flip-flop which was the source of the set signal.

**Ring Counter No. 2**

This ring counter is identical to the previous ring counter except one additional line has been added per stage. This line, labeled Z in the diagram, has been added to allow for overlapping of the positive pulses on the two clock lines. Aside from this allowable modification to the clock lines, the timing chart for this network is identical to that of ring counter No. 1.

*Logic Network (Two Stages Shown)*

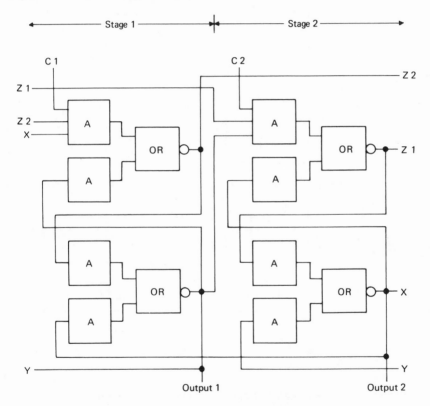

The additional line appears as two lines in this network, since two stages are shown. The Z line is connected so that it skips a stage before connecting to an AOI gate. This Z line is an inhibit line—a blocking line—in that it prevents the set condition from rippling through many stages when both clock lines are up.

**Ring Counter No. 3**

This ring counter is similar to ring counter No. 1 except it employs a different type of gated flip-flop.

The network operates on a two line clock system. The ring advances on positive pulse on either clock line. As with ring counter No. 1, both clock lines should not be up at the same time. The timing chart for this network is the same as for ring counter No. 1.

*Logic Network (Two Stages Shown)*

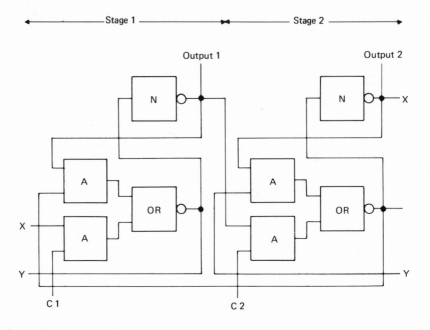

*Operational Notes*

The line labeled X is used to carry the set condition forward.

The line labeled Y is used to carry the reset signal back to the stage which has just passed its set condition forward.

# 5

# OR–INVERT, EMITTER COUPLED LOGIC

# OR-INVERT

## Use of Negative Logic

The OR-Invert logic gate is, as the name implies, an OR gate followed immediately by an inverter. Output signals cannot be taken from the OR but must be allowed to pass first through the inverter before being distributed to other gates. This restriction is a result of the circuit rules developed during the design of the gate and relates directly to the lack of voltage gain in the OR portion of the gate. Gates of this type are often called *NOR* gates, NOR standing for Not OR: an OR gate followed immediately by an inverter.

As was explained in Section 1 of this manual, the name assigned to a logic gate can vary with the definition of a "1" and "0." Throughout this text the author has consistently assumed the more positive voltage level has been assigned the symbol 1; and conversely, the more negative of the two voltage levels has been assigned the symbol 0 (this is called positive logic). But let us digress for a moment and assume we assign the 1 and 0 symbols the other way around (negative logic): a 0 being assigned to the higher level and a 1 to the lower. With this turnabout, an OR gate becomes an AND, and an AND gate becomes an OR gate. The NOR gate we are about to study becomes a NAND gate; conversely, a NAND becomes a NOR. Thus our problem with NOR gates can be easily dismissed by asking the user of this manual to switch to negative logic and refer to the NAND section. This is a valid procedure and is recommended by the author. On the other hand, a good designer should be able to work with NOR gates in positive logic. It is true that the designer should be able to obtain the same solution regardless of his direction of approach. Unfortunately, the art of logic design has not been reduced to a formal fixed procedure. It is in all senses of the word an art, and as such, the solutions will vary with the approaches. It is for this reason that the author has included in the manual a section on NOR logic.

## Product of Sums Solutions

In the NAND section of this manual a procedure for obtaining a sum of products solution from a Karnaugh map was discussed. This procedure involved drawing loops around clusters of 1's in the Karnaugh. These loops were then used to design a network whose output is up under every input combination that displays a 1 in the map. But suppose the 0's in the map, instead of the 1's, had been enclosed in loops; and these loops were then used to design a network. The function obtained in this manner is the complement of the original function that was plotted on the map. This procedure was found to be useful when working

217

with AOI gates and was discussed in Section 4. But let us examine this complemented function more fully. As the loops of 0's are read from the map they take the following form:

$$\overline{F} = A\,B + C\,D + \cdots$$

While the complemented function was useful for AOI gates, it is of little value for NOR gates. But the true function can be obtained from this equation by complementing both sides:

$$\overline{\overline{F}} = \overline{A\,B + C\,D + \cdots}$$
$$F = (\overline{A} + \overline{B})\,(\overline{C} + \overline{D})\,(\text{- - -})(\text{- - -})$$

The expression on the right of the equal sign is said to be in the P of S form; that is, *product of sums* form. When an expression of this type is implemented in AND and OR gates, it appears as a row of OR gates all connected to a common AND. Let us at this point work through a practice example.

*Example of a P of S Solution*

Given the following truth table:

| A | B | C | Output |
|---|---|---|--------|
| 0 | 0 | 0 | 1 |
| 0 | 0 | 1 | 1 |
| 0 | 1 | 0 | 1 |
| 0 | 1 | 1 | 0 |
| 1 | 0 | 0 | 0 |
| 1 | 0 | 1 | 1 |
| 1 | 1 | 0 | 0 |
| 1 | 1 | 1 | 0 |

Karnaugh mapping the above truth table and looping the 0's with a minimum set of loops produces Figure 5.1. Reading the loops of 0's from this map and translating them to an equation gives us the following:

$$\overline{F} = A\,\overline{C} + B\,C$$

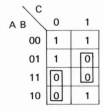

**Fig. 5.1** Karnaugh map with 0's looped

Inverting both sides of the equation:

$$\overline{\overline{F}} = A\,\overline{C} + B\,C$$
$$F = (\overline{A} + C)\,(\overline{B} + \overline{C})$$

The AND/OR implementation of this equation is shown in Figure 5.2. This type of solution is called a P of S implementation.

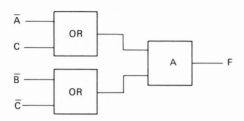

**Fig. 5.2**   Circuit implementation for $F = (\overline{A} + C)\,(\overline{B} + \overline{C})$

## Converting P of S to NOR

Product of sums expressions have particular significance in NOR logic design. The reason for this can be seen by examining Figure 5.3, which depicts a two level (signals pass through two successive gates) NOR network. As can be seen, the output expression of this network is in P of S form. Thus two levels of NOR gating appear, Booleanwise, as a row of OR gates all connected to a common AND. This characteristic enables a designer to move quickly from a Karnaugh map to a NOR network. There is, however, one notable exception in this conversion of OR/AND to NOR. If an input variable connects directly to the output AND, then this variable must be inverted before connecting to the output NOR. This problem is illustrated by implementing the following expression:

$$F = (A)\,(B + \overline{C})\,(\overline{D} + E)$$

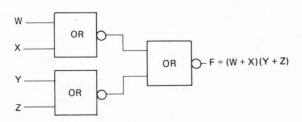

**Fig. 5.3**   Two level NOR network

If this expression is implemented in OR/AND gates, the A variable would connect to the output AND. But when this expression is implemented in NOR gates the A variable is inverted before connecting to the output NOR (see Figure 5.4).

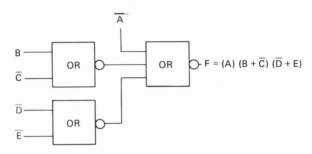

**Fig. 5.4** NOR network

In summary, the procedure for moving from a Karnaugh map to a network constructed of two stages of NOR gates is as follows:

1. Loop 0's on map.
2. Read Boolean expression for loops of 0's.
3. Invert Boolean expression to obtain a P of S expression.
4. Implement P of S expression in NOR gates, inverting single input variables that connect directly into the output NOR.

### Three Stage Gating

While two stage NOR networks are important, they usually require many input variables in true and complemented form. While input variables are often available in both forms, it is unwise to count on their availability. For this reason the author of this manual assumes the availability of the true variable only. Whenever a complemented variable is required, it is generated within the network. This procedure enables the user of this manual to connect the output of any network to the input of any other network. That is, the networks are consistent; they do not produce outputs in true and complemented form and do not require inputs in both forms. As we have seen in the NAND section of this manual, networks which do not use complemented input variables can require up to three stages of gating. The procedure for obtaining three stage gating was explained for NAND gates; and this same procedure, with some modifications, can be used with NOR gates. This modified method is explained by way of the following example:

Steps

Example problem

1. Draw Karnaugh map of desired function.

2. Mark square on map that indicates the *complements* of the available inputs. In this case we assume A and B are available so square $\overline{A}\,\overline{B}$ is marked.

3. All loops drawn on the map *must* include this marked square. Draw implicant loops on the map to cover all 0's. Be sure every loop covers the marked square, even if it must enclose some 1's.

4. These loops are used to construct a network where each loop requires one NOR gate. One additional NOR gate is required for the output. The input line labels to the left row of NORs are obtained by referring to loops on the map. Each loop is identified by a variable or a series of variables, and these variables are *inverted* before being used as line labels. Loop $\overline{A}$ becomes A, and so on.

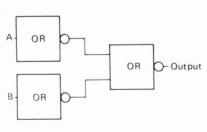

5. At this point the network is correct except for the 1's that have been enclosed in loops of 0's. These 1's must be blocked or inhibited. This is accomplished by considering these 1's as a new problem and mapping them as 0's on a new map.

6. Again, draw loops on the map to enclose all 0's. Be sure every loop encloses the marked square. It should not be necessary to enclose any 1's at this point.

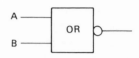

7. A NOR gate is drawn for each of these loops. The input labels to these gates are obtained by complementing each of the variables obtained from the loop labels. Loop $\overline{A}\,\overline{B}$ becomes inputs A and B to the same gate.

8. The outputs of these gates may be thought of as inhibit terms. Each NOR is related to a specific 1 or set of 1's that was previously enclosed. The outputs of these NORs are selectively connected to the previously designed NORs, making sure that each undesired 1 is inhibited from each loop. If a 1 is enclosed by more than one loop, it must be inhibited from each enclosing loop.

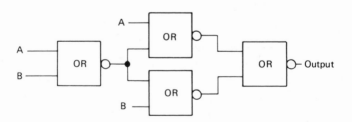

The above procedure will always produce a minimum three stage NOR network. When the original problem has a 0 in the square labeled with all true variables, then the first looping must enclose the entire map. When this occurs, the above procedure produces a network with a single input NOR (inverter only) as an output gate.

## EMITTER COUPLED LOGIC

### General Information

The *emitter coupled logic* (ECL) networks in this manual differ from other networks in that they relate to a specific circuit. This fact should be obvious from the title, emitter coupled logic, which is the name of a circuit rather than the name of a logic connective. As a matter of fact, the author has gone to great lengths in other sections of this manual to avoid relating the listed networks to specific circuits. Relating each network to a specific circuit unnecessarily narrows the usefulness of the manual. At the present time there are mechanical, hydraulic, and numerous electronic NAND gates which can be used to implement the networks listed in the AND-Invert section; and narrowing this section to a specific circuit would serve no useful purpose. But the ECL circuit is widely used and is so different from other logic connectives that it cannot benefit from their designs. For these reasons the author has made an exception of the ECL circuit and has devoted many pages to networks which are specifically implemented with this gate. Possibly other circuits or mechanical devices will warrant this treatment in the near future.

### Emitter Coupled Logic Circuit

The basic emitter coupled logic circuit (ECL) is shown in Figure 5.5.

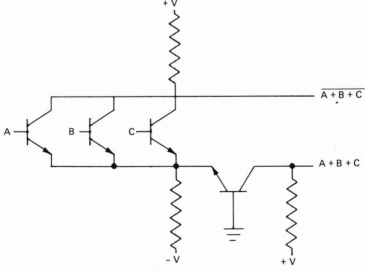

**Fig. 5.5** ECL circuit

The outstanding characteristic of this circuit is that it has two outputs that are the complements of each other. The lower output in Figure 5.5 produces the OR function of all inputs, and the upper output is the inverted OR (NOR) function.

In addition to the advantage of having a true and a complemented output, this circuit has one more desirable characteristic. The lower outputs of two circuits can be connected together to form a wired AND. This AND function is very useful, as it immediately follows an OR gate. As we have seen, OR/AND networks can be read directly from a Karnaugh map. These OR/wired AND networks find many applications in sequential networks. The wired AND makes the appearance of these networks somewhat surprising, as is demonstrated in Figure 5.6(a). Until one becomes proficient at reading these networks, it may be advisable to draw a small AND gate at the point of the wired AND, as shown in Figure 5.6(b).

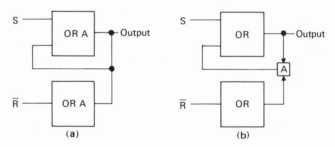

**Fig. 5.6**  ECL network showing usage of wire AND

In some cases the complemented outputs of two ECL gates may also be wire ANDed together, as shown in Figure 5.7(a). The wire ANDing of these outputs together serves only to extend the total number of acceptable inputs. While this fan-in extension is important, it is not used in this manual for the reason that no limitation has been placed on the fan-in. The author assumes that the designer will use this technique if it is needed and is available.

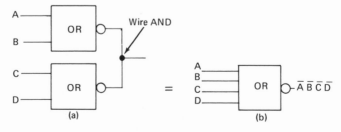

**Fig. 5.7**  ECL network showing wire AND on complemented outputs

## Using the ECL Gate

Since one of the outputs of the ECL gate produces the NOR function, this circuit may be used as a NOR gate or, for that matter, a NAND gate in negative logic. In many cases the NOR or NAND approach is the best design technique. However, when many two stage networks are required it is usually advisable to change to OR/AND logic, thus taking advantage of the wired AND. This approach is excellent for many of the flip-flop designs, as this manual points out. This OR/AND approach has one severe weakness that limits its applications. The two or more functions that are wire ANDed together cannot be used separately as subfunctions for other networks. Wired connectives do not give isolation to the incoming lines and thus are less useful than true connectives.

Choosing between the NOR approach and the OR/AND approach is a minor problem, for two stages of NOR gates are equilvalent to OR/AND. This makes the conversion from NOR to OR/AND a simple matter. The author therefore suggests a NOR design followed by a selective conversion to OR/AND. The conversion takes place only where subfunctions are not required and an overall advantage is obtained.

## OR-INVERT AND EMITTER COUPLED LOGIC
## COMBINATIONAL NETWORKS

**Exclusive-OR**

Inputs are required in true and complemented form for this network. The output of this network is up when one and only one input is up.

*Output Expression*

Output = A $\overline{B}$ + $\overline{A}$ B

*Logic Network*

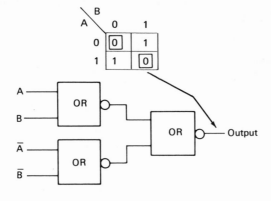

**Exclusive-OR Complement No. 1**

Input complements are not required for this network, but the output signal is delayed by three OR-Invert gates.

The output of this network is up when no inputs or both inputs are up.

*Output Expression*

Output = A B + $\overline{A}$ $\overline{B}$

*Logic Network*

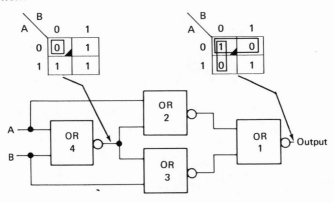

*Design Note*

Notice how gate 4 is used to block or inhibit the upper left-hand 1 on the final output map.

### Exclusive-OR Complement No. 2

Input complements are not required for this network which makes use of a wired AND connective.

*Output Expression*

Output = A B + $\overline{A}\ \overline{B}$

*Logic Network*

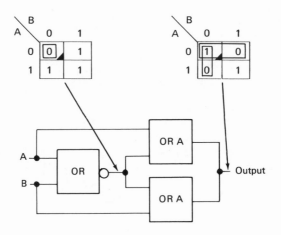

*Design Note*

Notice how the mapping for this implementation is the same as that for the preceding OR-Invert network.

### Even Circuit—Three Input

The output of this circuit will be up when all inputs are down or any two inputs are up and the third is down.

*Output Expression*

$$\text{Output} = \overline{A}\,\overline{B}\,\overline{C} + A\,B\,\overline{C} + A\,\overline{B}\,C + \overline{A}\,B\,C$$

*Logic Network*

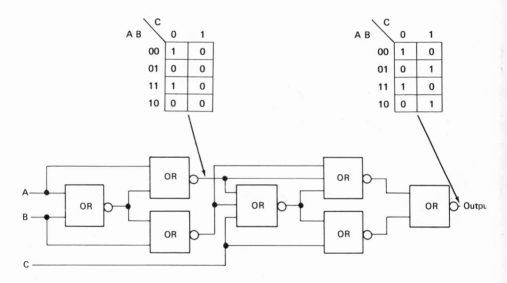

**Odd Circuit—Three Input**

The output of this circuit will be up when one input or three inputs are up.

*Output Expression*

Output = A B C + A $\overline{B}$ $\overline{C}$ + $\overline{A}$ B $\overline{C}$ + $\overline{A}$ $\overline{B}$ C

*Logic Network*

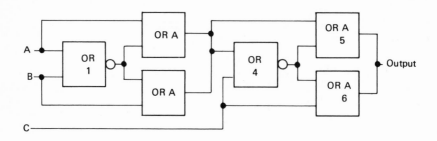

*Design Note*

This network was designed by cascading two Exclusive-OR complement networks.

### Even Circuit—Four Input No. 1

The output of this circuit will be up if no inputs, two inputs, or four inputs are up.

Output = $\overline{A}\ \overline{B}\ \overline{C}\ \overline{D} + \overline{A}\ \overline{B}\ C\ D + \overline{A}\ B\ \overline{C}\ D + \overline{A}\ B\ C\ \overline{D} + A\ \overline{B}\ \overline{C}\ D$

$+ A\ \overline{B}\ C\ \overline{D} + A\ B\ \overline{C}\ \overline{D} + A\ B\ C\ D$

*Logic Network*

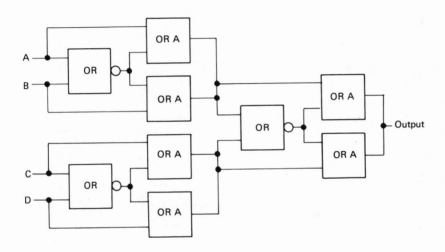

*Design Note*

This network was designed by cascading Exclusive-OR complement networks.

**Even Circuit—Four Input No. 2**

The output of this circuit will be up if no inputs, two inputs, or four inputs are up.

*Output Expression*

$$Output = \overline{A}\,\overline{B}\,\overline{C}\,\overline{D} + \overline{A}\,\overline{B}\,C\,D + \overline{A}\,B\,\overline{C}\,D + \overline{A}\,B\,C\,\overline{D} +$$
$$A\,\overline{B}\,\overline{C}\,D + A\,\overline{B}\,C\,\overline{D} + A\,B\,\overline{C}\,\overline{D} + A\,B\,C\,D$$

*Logic Network*

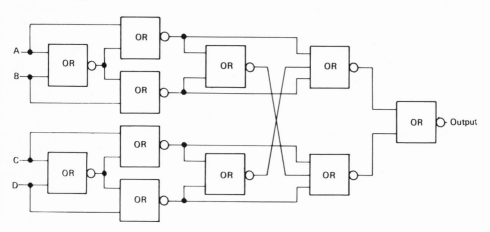

### Decoder—Two Variable

Minimum NOR gate solution under the assumption that no input complements are available.

#### *Operation*

One and only one of the output lines will be up for each of the four possible input conditions.

#### *Karnaugh Mappings*

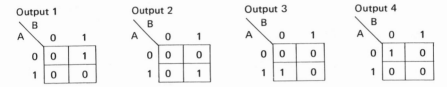

Output 1

| A \ B | 0 | 1 |
|---|---|---|
| 0 | 0 | 1 |
| 1 | 0 | 0 |

Output 2

| A \ B | 0 | 1 |
|---|---|---|
| 0 | 0 | 0 |
| 1 | 0 | 1 |

Output 3

| A \ B | 0 | 1 |
|---|---|---|
| 0 | 0 | 0 |
| 1 | 1 | 0 |

Output 4

| A \ B | 0 | 1 |
|---|---|---|
| 0 | 1 | 0 |
| 1 | 0 | 0 |

#### *Logic Network*

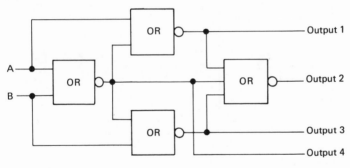

**Adder—Half**

A half adder design based on the Exclusive-OR complement network. Inputs to this network are required solely in complemented form, and likewise, the sum output line is inverted.

*Output Expressions*

$$\overline{\text{Sum}} \quad = \text{X Y} + \overline{\text{X}}\,\overline{\text{Y}}$$

$$\text{Carry} \quad = \text{X Y}$$

*Logic Network (OR-Invert)*

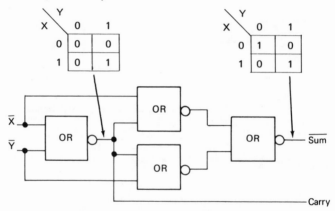

*Logic Network (ECL)*

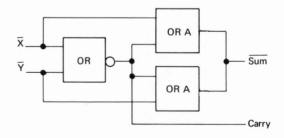

## Adder—Full No. 1

Products of sums solution.    Input complements required for all variables.

*Karnaugh Mappings*

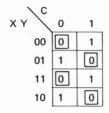

Sum $= (X + Y + C)\ (X + \bar{Y} + \bar{C})\ (\bar{X} + \bar{Y} + C)\ (\bar{X} + Y + \bar{C})$
Carry $= (X + C)\ (X + Y)\ (Y + C)$

*Logic Network (OR-Invert)*

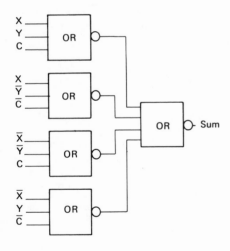

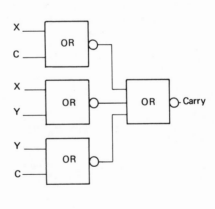

*Logic Network (ECL)*

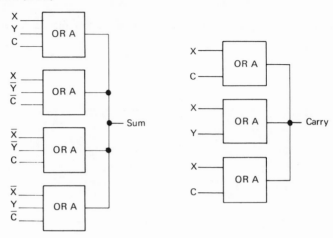

## Restriction

This particular set of networks is of little practical value, since all input variables are required in true and complemented form. In addition, these networks require more than a minimum set of gates.

**Adder—Full No. 2**

Minimum NOR gate solution for condition of no input complements available.

Signal delay from carry-in to carry-out is two gates.

*Output Expressions*

Sum = $\overline{X}\,\overline{Y}\,\overline{C} + \overline{X}\,Y\,C + X\,Y\,\overline{C} + X\,\overline{Y}\,C$

Carry = $X\,Y + Y\,C + X\,C$

Inverted Propagate = $\overline{X + Y}$

Logic Network

238

## Adder—Full No. 3

The following adder was designed by putting together two Exclusive-OR complement networks. The Exclusive-OR complement networks selected make use of a wired AND connective.

*Output Expression*

$$\text{Sum} = X \, \overline{Y} \, \overline{C} + \overline{X} \, Y \, \overline{C} + \overline{X} \, \overline{Y} \, C + X \, Y \, C$$
$$\text{Carry} = X \, Y + Y \, C + X \, C$$

*Logic Network*

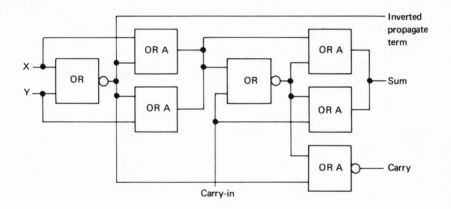

## Adder Speed-up Circuit No. 1

*Basic Principle*

The speed of a functional adder is determined by the rate of propagation of the carry information. Most full adder implementations require the carry-in signal to pass through two levels of logic before appearing as a carry-out signal. This delay in the carry ripple can be reduced by using the following circuit or speed-up circuit No. 2. Both of these speed-up circuits require an inverted propagate signal from each full adder position. The Boolean expression for this inverted propagate term can be either $\overline{A + B}$ or $A B + \overline{A} \, \overline{B}$.

*Logic Network*

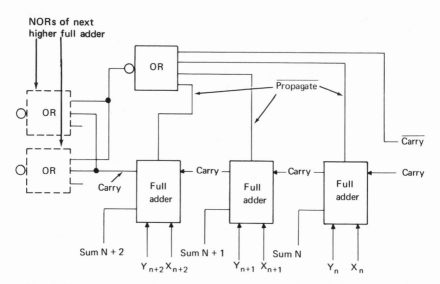

The NOR gates shown in dotted lines are actually part of the next full adder position and do not represent additional gates. Notice that the output of the speed-up NOR gate connects as an input to all NOR gates that are fed from the normal carry-out.

The inverted carry-in signal will require an inverted gate for its generation if NOR logic is being used. With ECL, this signal is available from the alternate output terminal of the lower order carry generation gate.

*Extension*

The bridging of the carry signal can be expanded to any number of stages as long as each bypassed stage furnishes an inverted propagate signal to the speed-up NOR.

## Adder Speed-Up Circuit No. 2

This circuit is the same as speed-up circuit No. 1 except that the same principle has been applied for a second time.

*Logic Network*

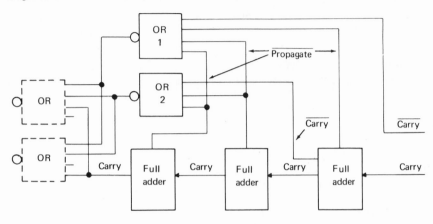

*Operational Notes*

The function of NOR gate 1 has been explained in the preceding circuit. The output of gate 2 is connected, as is gate 1, to all NOR carry-in gates of the next higher full adder position. The function of gate 2 is to speed up a carry signal that is generated in the rightmost full adder.

*General Principle*

The speed-up technique as shown in this and the preceding circuit can be applied and reapplied. The bridges can be made as long as desired (within fan-in, fan-out limitations), and numerous levels of bridging can be used. The number of levels and the length of the bridges that can be effectively used are a function of the length of the adder and the fan-in/fan-out characteristics of the logic gates being used. For this reason no general rules can be given, but it is not uncommon for the bridging to require as many logic gates as are required by the basic full adders themselves.

## OR-INVERT AND EMITTER COUPLED LOGIC
## SEQUENTIAL NETWORKS

### Flip-Flop, Set/Reset Type No. 1 (Flip-Flop Latch)

The following network is the basic flip-flop for OR-Invert logic. It has two stable states when both inputs are down.

*Flow Chart*

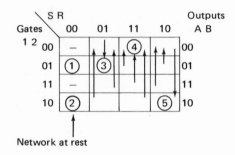

Network at rest

*Logic Network*

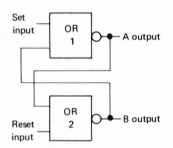

*Operational Notes*

Both inputs are normally down.

Raising the set input will set the network ($A = 1$, $B = 0$), while raising the reset input will reset the network ($A = 0$, $B - 1$).

The two output lines are complements of each other except when both inputs are up, in which case both outputs are 0.

*Additional Inputs*

Additional set lines may be connected as inputs to NOR 1. These lines should be down when not in use, and the raising of any set line will result in setting the flip-flop.

## Flip-Flop, Set/Reset Type No. 2

A basic two state memory element which makes use of a wired AND connective.

*Flow Chart*

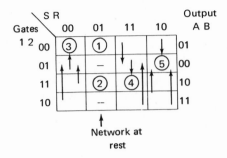

Network at
rest

*Logic Network*

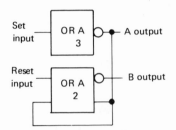

The wired AND gate is considered gate 1 in the above flow chart.

*Operational Notes*

The reset input is normally up and the set line is normally down.

Raising the set line will set the flip-flop (A = 1, B = 0), while lowering the reset input will reset the flip-flop (A = 0, B = 1).

*Additional Input*

Additional set lines may be connected as inputs to gate 2. These inputs should be down when not in use, and the raising of any set line will result in setting the flip-flop.

### Inverter, Nonoverlapping

This network generates a true and an inverted signal which are nonoverlapping in the up level; both outputs are never up at the same time regardless of input pattern.

Networks of this type are particularly useful for driving shift registers and ring counters.

*Input-Output Timing Chart*

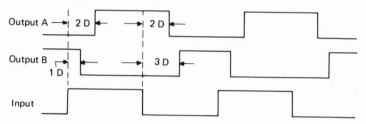

D = delay through one NOR gate

*Flow Chart*

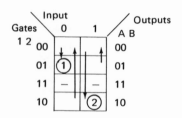

*Logic Network*

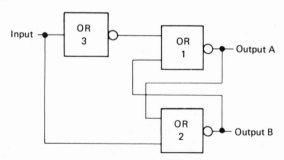

*Operational Notes*

As the input is raised and lowered the network moves from stable state ① to ② and back to ① . But the operating point does not move directly to its next state; it moves first through an unstable state where both outputs are down. This technique prevents both outputs from ever being up at the same time.

## Flip-Flop, Gated No. 1

A flip-flop with one or more inputs where each input is under control of a separate gate line. When the network is at rest the reset line is down and the gate line (lines) are up.

*Partial Flow Chart*

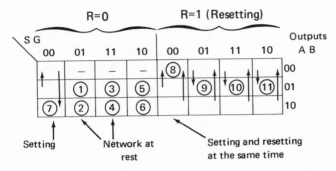

*Logic Network*

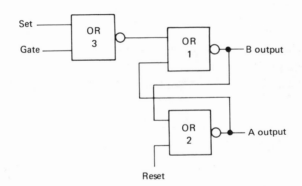

*Operational Notes*

Reset line is normally down.

Gate line is normally up.

The network is reset ($A = 0$, $B = 1$) by raising the reset line.

The network is set ($A = 1$, $B = 0$) by lowering the gate line when the set line is down. If the gate line is lowered when the set line is up, it will have no effect on the network.

A second gated set line may be added to the network by duplicating NOR No. 3. In this way the lowering of either gate line, when its associated set line is down, will result in setting the flip-flop.

### Flip-Flop, Gated No. 2

A basic two state memory element with gated set lines. The operation of this network is the same as gated flip-flop No. 1 except that the reset line is up instead of normally down. A second gated set line is shown on this network.

*Logic Network*

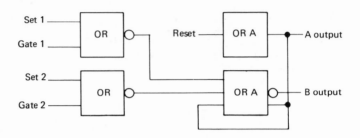

*Operational Notes*

Reset line is normally up.

Lowering the reset line will reset the flip-flop ($A = 0$, $B = 1$).

Gate lines are normally up.

Lowering a gate line when its associated set line is down will result in setting the flip-flop ($A = 1$, $B = 0$).

## Flip-Flop, Automatic Reset Type No. 1

This network requires no reset line, since the value of the data line (either 1 or 0) is entered into the flip-flop when the gate line is lowered.

*Partial Flow Chart*

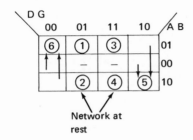

Network at rest

*Logic Network*

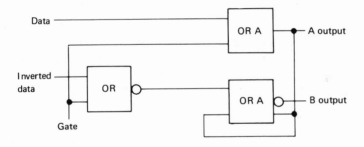

*Operational Notes*

Gate line is normally up.

When the gate line is lowered the value of the data line is entered into the flip-flop.

As long as the gate line is down, the flip-flop will follow the value of the data line.

### Flip-Flop, Automatic Reset Type No. 2

This flip-flop, unlike the previous network, requires no inverted data line. The flip-flop moves to the value of the data line when the gate line is lowered.

*Flow Chart*

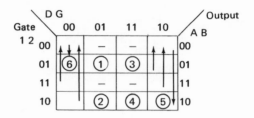

*Logic Network*

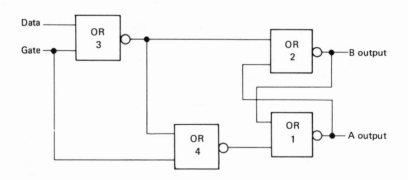

*Operational Notes*

Gate line is normally up.

Lowering the gate line will result in moving the flip-flop to the value of the data line.

When the gate line is up, the flip-flop is independent of changes on the data line.

## Double Gating

This particular flip-flop is well suited to applications requiring more than one source of data. In applications of this type the gating signals are used to select the input source and the clock signal determines the timing. Gate lines and the clock line are normally up.

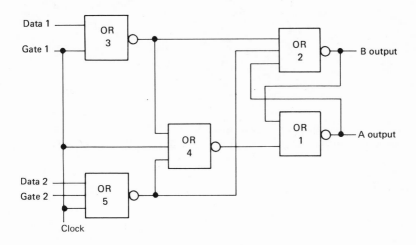

### Flip-Flop, Automatic Reset Type No. 3

This network is based on a modification of auto flip-flop No. 1. This modification enables the flip-flop to operate without a complemented data line.

*Flow Chart*

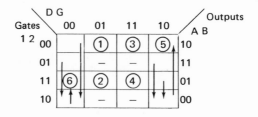

Gate 2 is the output of the wired AND.
Gate 1 is the OR output of bottom right-hand gate before wire ANDing.

*Logic Network*

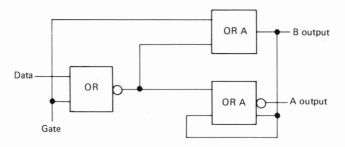

*Operational Notes*

Gate line is normally up.
Lowering the gate line will result in moving the flip-flop to the value of the data line.
When the gate line is up, the flip-flop is insensitive to changes on the data line.

### Design Note

By comparing this network with auto reset flip flop No. 1, it can be observed that the input line that has actually been saved by this network is the data line. The inverted data from No. 1 is labeled *Data* in the above network. This reversal of input labels is compensated for by relabeling the output lines.

### Double Gating

This network can be modified to accommodate double gating, as explained in auto reset flip-flop No. 2.

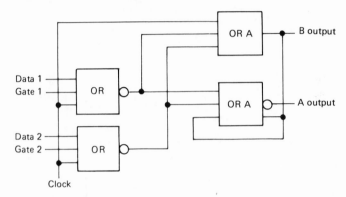

This double gated network will now accept data from a data line that has its associated gate line down, but only when the clock is down.

## Flip-Flop, Three State

This network is the natural extension of the two state flip-flop. As with the two state flip-flop, all NOR gates feed all other NOR gates.

This network has three stable states when all inputs are down.

*Partial Flow Chart*

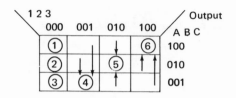

*Logic Network*

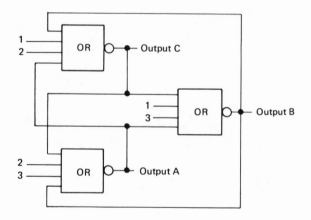

*Operational Notes*

All inputs are normally down.

One and only one output is up when the network is at rest.

Raising an input line will result in raising a corresponding output line.

### Flip-Flop, T Type No. 1

A flip-flop whose output changes each time the input line is lowered. Raising the input line has no effect on the output. Flip-flops of this type are used in the construction of binary counters.

*Partial Flow Chart*

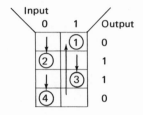

*Timing Chart*

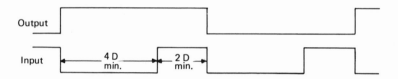

*Logic Network*

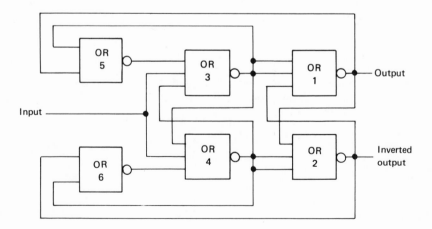

*Complete Timing Chart of All Gates*

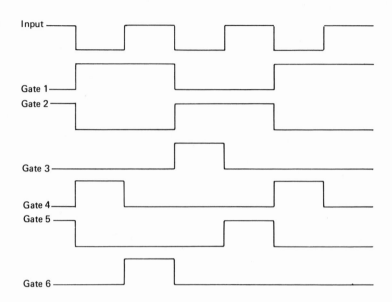

### Design Note

This network is in effect a four stage ring counter. This fact can be observed by examining gate outputs 4, 6, 3, 5 in the complete timing chart.

### Additional Inputs

Normally up set and reset lines may be added to gates 6 and 5, respectively. The network should be set or reset only when the input line is up.

**Flip-Flop, T Type No. 2**

A T type flip-flop that changes state on the fall of the input line.
This flip-flop is preferred over the previous flip-flop for the following reasons:

1. Network can operate at a maximum rep-rate with a symmetrical input pulse pattern.
2. Network can be set or reset with input line up or down.

*Flow Chart*

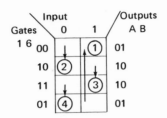

*Timing Chart*

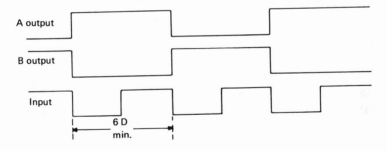

## Logic Network

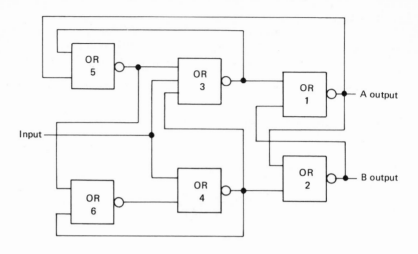

## Additional Input

This flip-flop can be set in an output condition of A = 1, B = 0 or reset to an output condition of A = 0, B = 1 by adding lines as indicated in the following table.

| Condition of input when setting or resetting is to occur | Desired output condition | Add an input line to the following gates |
|---|---|---|
| Input line down | A = 1, B = 0 | Gate 6 |
| | A = 0, B = 1 | Gates 4 and 5 |
| Input line up | A = 1, B = 0 | Gate 2 |
| | A = 0, B = 1 | Gate 1 |
| Input unknown | A = 1, B = 0 | Gates 6 and 2 |
| | A = 0, B = 1 | Gates 1, 4, and 5 |

All added lines must be maintained at the 0 level when not being used.

### Flip-Flop, T Type No. 3

A flip-flop whose output changes each time the input line is raised.

This network is identical to T flip-flop No. 2 except the output is taken from gate 6 rather than gate 1.

*Flow Chart*

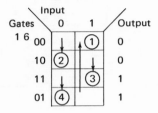

*Timing Chart*

Since this network is used as a building block for more complex networks, it is advisable to examine a timing chart which covers all 6 gates.

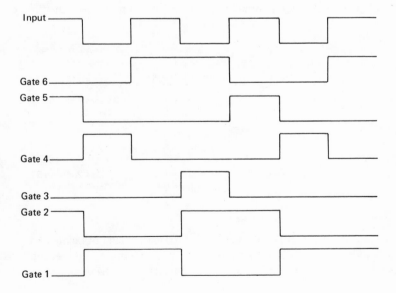

*Logic Network*

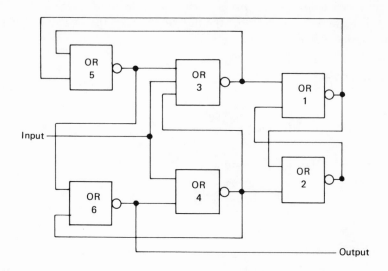

*Additional Inputs*

Use the following table for adding lines which will set or reset the flip-flop.

| Condition of input when setting or resetting is to occur | Desired output condition (gate 6) | Add an input line to the following gates |
|---|---|---|
| Input line down | Output = 1<br>Output = 0 | Gates 4 and 5<br>Gate 6 |
| Input line up | Output = 1<br>Output = 0 | Gate 2<br>Gate 1 |
| Input unknown | Output = 1<br>Output = 0 | Gates 2, 4, and 5<br>Gates 1 and 6 |

All added lines must be maintained at the 0 level when not being used.

An additional T input may be added to this network by connecting it as an input to gates 3 and 4. In this case only one input should be up at a time.

## Flip-Flop, Set Dominant

A set dominant flip-flop is a two state network having a set and a reset input. It operates as a conventional set/reset flip-flop except when simultaneous set and reset signals are applied. The set dominant flip-flop will set under this input condition and remain set regardless of the order of removal of input signals.

*Flow Chart*

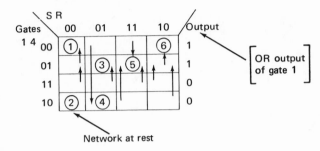

Network at rest

*Logic Network*

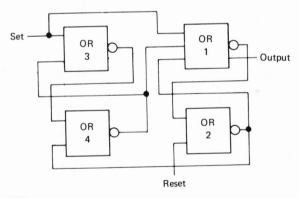

Reset

*Operational Notes*

Both inputs are normally down.
Raising the set input will set the flip-flop (output = 1).
Raising the reset input will reset the flip-flop (output =0).
Raising both inputs will result in setting the flip-flop, and the order of raising the inputs will have no effect on the network (see flow chart).

*Additional Inputs*

Additional set lines may be connected to gates 1 and 3. The raising of any set

## Flip-Flop, Post Indicating

This flip-flop differs from a set/reset flip-flop in that the output does not change until the input signal (set or reset) has been applied and then removed.

*Flow Chart*

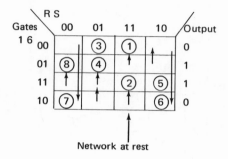

Network at rest

*Logic Network*

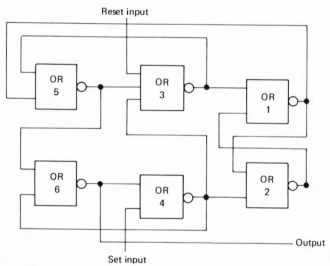

*Operational Notes*

Both inputs are normally up.

Lowering the set input will set the flip-flop, but the output will not change until the reset line is returned to its normally up level.

Lowering the reset input will reset the flip-flop, but the output will not change until the reset line is returned to its normally up level.

If both inputs are lowered at the same time, the output will not change; but the final output level will be determined by the input line that is raised last (see flow chart).

## Sampling Gate

This network is sometimes called a picture-taking flip-flop. The value of the data line is frozen in the flip-flop when the gate line is dropping. In effect, the network takes a picture of the data line as the gate line falls.

*Flow Chart*

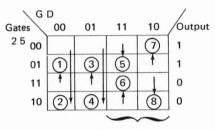

Network at rest

*Logic Network*

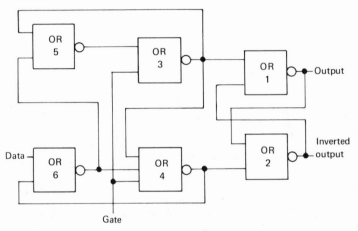

*Operational Notes*

The gate line is normally up, and the flip-flop is insensitive to changes on the data line.

On the fall of the gate line, the value of the data line is entered into the flip-flop. The value stored in the flip-flop and the output lines will now remain stable until the gate line is raised and again lowered.

*Design Note*

Logic gates 1 and 2 may be removed from this network and the output taken from logic gate 4. When this modification is made, the sample taken by the network will be held only as long as the gate line is down. The output (gate 4) will be down at all times when the gate line is up.

## One Shot Circuit (Single Shot)

When the input line to this network is raised, the output line will supply a plus-going pulse of constant time duration. The input line can be returned to its normally down level at any time without affecting the output pulse duration. The output pulse duration is determined by the delay element.

*Logic Network*

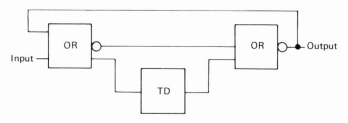

*Timing Chart*

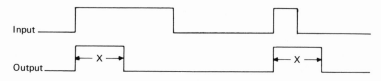

X = the time delay imposed by the delay element

*Restriction*

After generating an output pulse, the network cannot be recycled until the delay path has been cleared. This requires a time interval of X units.

## Binary Counter No. 1

The following network has been designed to count the number of positive pulses appearing on a line. The count is continually displayed in binary, one output line per T flip-flop. This network is the simplest of all the counters.

*Logic Network*

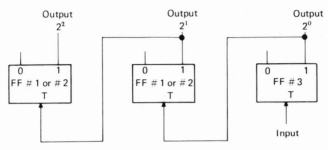

All flip-flops are T type

*Timing Chart*

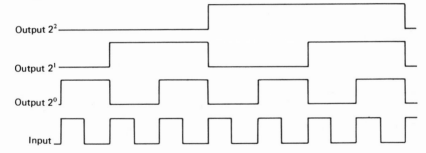

*Design Notes*

The right-hand flip-flop is of the type that changes output on the rise of the input line, while all other flip-flops change outputs on the fall of the input line.

It should be noted that inputs to all but the low order flip-flop may be up when the counter is at rest between counts. This can introduce problems when it comes to setting the network to a precount value. This is especially true for flip-flop type No. 1 which cannot be set or reset when its input line is up.

*Design Alternative*

If the right-hand flip-flop is replaced with a T type flip-flop that changes output on the fall of its input line, then the network will count the number of negative pulses on the input line.

## Binary Counter No. 2

The following binary counter is strongly suggested for most counting applications. It has two very desirable characteristics:

1. The stored count can be increased by any power of 2.
2. The carry lines between adjacent stages are all at the down level when the counter is not being stepped. This characteristic enables the counter to be easily preset to any value.

*Logic Network*

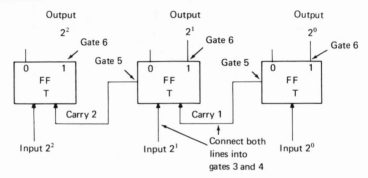

All flip-flops are T type No. 3

*Timing Chart*

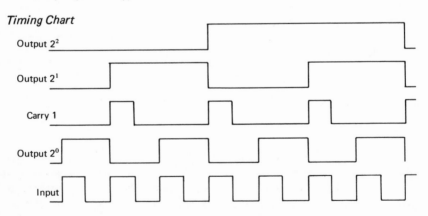

*Operational Notes*

The stored count will be increased by one each time the $2^0$ input is raised. The stored count will be increased by $2^1$ each time the $2^1$ input is raised, and so on.

All input lines should be maintained at a down level when not being used to step the counter. This requirement is a result of the ORing together of the carry input and the count input within each flip-flop.

## Counter, Gray Code

The following network will count the number of plus-going pulses appearing on the input line and display the output in Gray code. Only three stages are shown, but more stages may be added to the center section. The first and last stages are unique in their external connections.

*Logic Network*

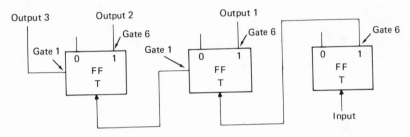

All flip-flops are T type No. 3

*Timing Chart*

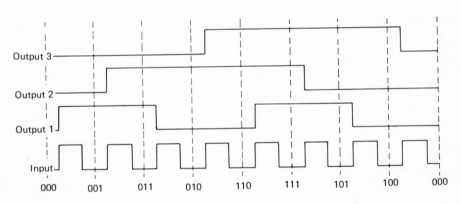

*Design Note*

The output from gate 1 of T flip-flop No. 3 changes on the fall of the input signal to that flip-flop.

## Shift Cell No. 1

The following network is a basic shift cell requiring a four clock drive system.

*Logic Network*

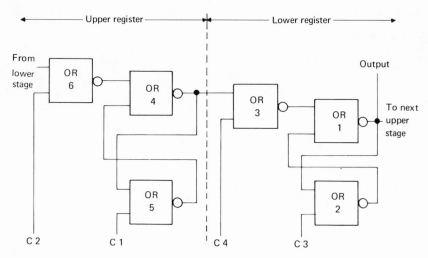

*Clocking System*

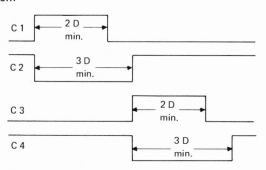

Under no circumstances should C 2 and C 4 be down at the same time.

*Operational Notes*

The raising of C 1 sets the upper register to all 1's.
The lowering of C 2 gates data from the lower register into the upper register.
The raising of C 3 sets the lower register to all 1's.
The lowering of C 4 gates data from the upper register into the lower register.

*Additional Inputs*

Data may be entered directly into the lower register by adding set and reset lines to gates 2 and 1, respectively. These lines should be down when not in use.

*Multimode Shifting*

If shifting in more than one mode is required, gate 6 should be duplicated.

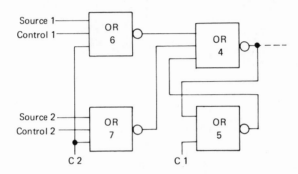

With this multimode arrangement the shift cell will now accept its next value from the source line which has its corresponding control line down. The control lines should be changed only when C 2 is up.

## Shift Cell No. 2

This shift cell is similar to shift cell No. 1 except that it makes use of a wired AND connective. This network has the advantage of requiring only negative-going clock pulses.

*Logic Network*

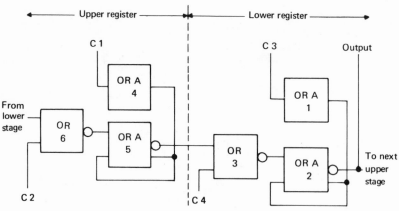

*Clocking Systems*

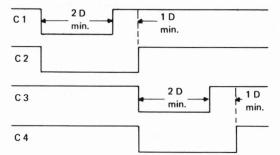

Under no circumstances should C 2 and C 4 be down at the same time.

*Operational Notes*

    The lowering of C 1 sets the upper register to all 1's.
    The lowering of C 2 gates data into the upper register.
    The lowering of C 3 sets the lower register to all 1's.
    The lowering of C 4 gates data into the upper register.

*Multimode Operation*

For multimode operation gate 6 must be duplicated, as in shift cell No. 1.

## Shift Cell No. 3

The following network is a two clock system shift cell. It has simplicity in design and operation.

*Logic Network*

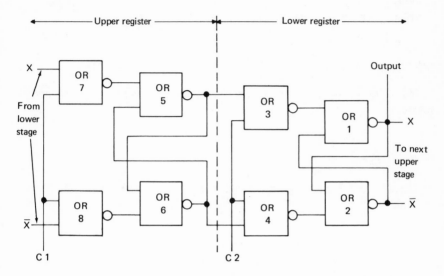

*Clocking System*

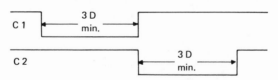

Under no circumstances should both clock lines be down at the same time.

### Operational Notes

Note that the two lines connecting together adjacent cells are actually complements of each other. This enables the cells to operate without reset lines.

When C 1 is lowered, data is gated into the upper register. Both true and complemented lines are gated, driving the upper register to a new value without resetting.

When C 2 is lowered, data is gated into the lower register.

### Additional Inputs

Normally down set and reset lines may be added to gates 2 and 1, respectively.

### Multimode Operation

Gates 7 and 8 must be duplicated if shifting in more than one mode is required.

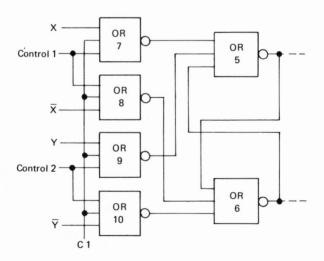

The multimode modification enables the shift cell to accept data from either of two sources, depending on which control line is down. The control lines should be changed only when C 1 is up if minimum clock pulsing is being used.

## Shift Cell No. 4

Like shift cell No. 3 this network requires a two clock drive system, but unlike No. 3 this network does not require a complemented data input line.

*Logic Network*

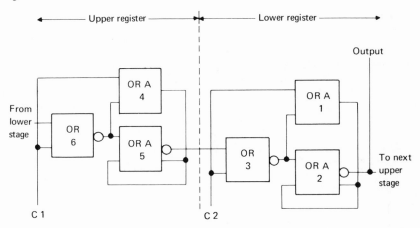

*Clocking System*

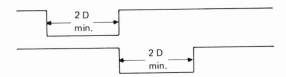

*Operational Notes*

> When C 1 is lowered, data is gated into the upper register.
> When C 2 is lowered, data is gated into the lower register.

*Additional Inputs*

A normally down reset line may be added to gate 2. An additional OR gate can also be added to the wired AND of gates 1 and 2 to give a normally up set line for entering data into the lower register.

## Multimode Operation

Gate 6 must be duplicated if shifting in more than one mode is required.

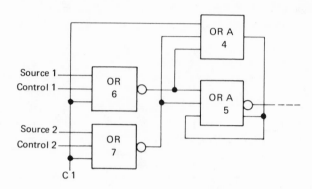

With this multimode modification, the network will now accept data from the source line whose associated control line is down. The control lines should be changed only when C 1 is up.

### Shift Cell No. 5

A six gate shift cell requiring only one clock drive line. This network and the following shift cell were obtained by a formal sequential design procedure. While this procedure is very effective, it produces networks that are difficult to partition into functional units such as upper and lower registers. To obtain a working understanding of this network, one should carefully study the flow chart and the operational notes.

*Flow Chart*

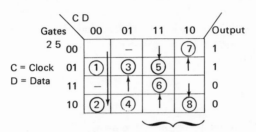

Network resides in either of these two columns when not being stepped.

*Logic Network*

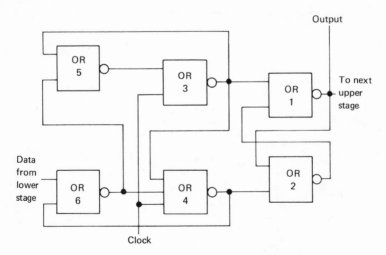

*Clock System*

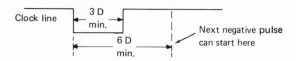

*Operational Notes*

When the clock line is up, the output of the network is insensitive to changes on the data line.

When the clock line goes down, the network samples the data line and adjusts the output to match. After sampling the data line the network becomes insensitive to further changes on the data line. Proper operation of the entire shift register is based on the fact that this cell becomes insensitive to changes in its input before its output changes.

*Additional Inputs*

Set and reset inputs may be added to gates 2 and 1, respectively. These inputs should be down when not in use.

*Multimode Shifting*

Gate 6 must be duplicated if shifting in more than one mode is required.

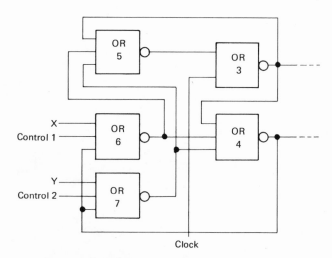

The shift cell will accept data from the X input if control line 1 is down or from the Y input if control line 2 is down.

## Shift Cell No. 6

This shift cell was designed from the same flow chart as that for shift cell No. 5 and therefore requires only one clock drive line. This network can be operated at a slightly higher rep-rate than cell No. 5, but it does require two interconnecting data lines between stages.

*Flow Chart*

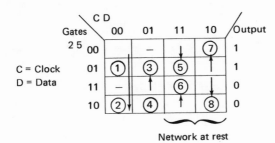

C = Clock
D = Data

Network at rest

*Logic Network*

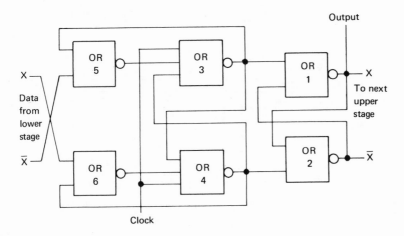

*Clock System*

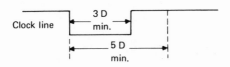

### Operational Notes

The operational notes for this shift cell are the same as for cell No. 5. Notice that each cell requires the passing of true and complemented data between adjacent cells.

### Additional Inputs

Normally down set and reset lines may be added to gates 2 and 1, respectively.

### Multimode Shifting

Gates 5 and 6 must be duplicated if shifting in more than one mode is required.

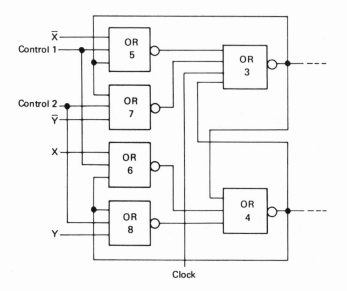

The shift cell will accept data from the X inputs if control line 1 is down or from the Y inputs if control line 2 is down.

## Ring Circuit No. 1

The following ring network requires a two clock drive system for advancing the ring. The two clock lines are alternately lowered then raised, advancing the ring 1 step for each complete pulse on the alternate drive lines.

This network has been designed by interconnecting a series of gated flip-flops.

*Logic Network (Two Stages Shown)*

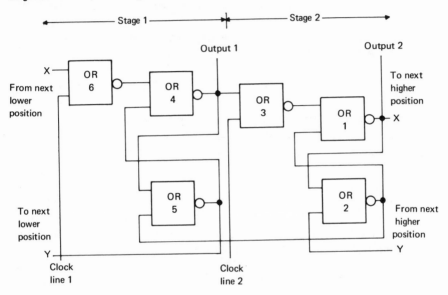

*Timing Chart (Four Stages Shown)*

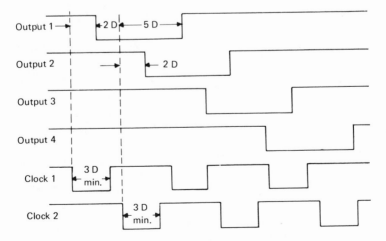

## Operational Notes

The outputs of this ring are normally up, and a down condition is propagate around the ring.

When clock line 1 is lowered, all odd number stages are conditioned to accept a down condition from their next lower neighbors. If the next lower neighbor is in an up position, then the accepting stage remains in an up position. If the next lower neighbor is in a down condition, then the accepting stage moves to a down state and sends a set signal back to the lower stage. This set signal causes the lower stage to move to an up position.

When clock line 2 is lowered, all even number stages operate in the same manner as described above.

## Restrictions

Under no circumstances should both clock lines be down at the same time.

Network will not maintain down levels in adjacent stages.

This network can be used only when an even number of stages is tolerated.

## Additional Inputs

All positions of the ring can be set to their normally up levels by adding normally down inputs to gates 2 and 5. A down level may be set into the ring by adding normally down inputs to gates 1 and 4.

## Ring Circuit No. 2

This network shows a modification that can be made in ring circuit No. 1 to enable it to operate on a two clock system which has overlapping down levels.

*Logic Network*

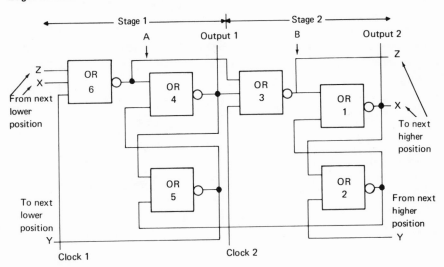

*Timing Chart*

Only the two clock lines are shown to point out the overlap problem which has been solved.

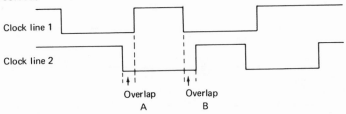

*Operational Notes*

The problem with overlapping clock pulses is that all stages will accept a down level when both clock lines are down. A down level will, therefore, advance at an uncontrolled rate, not stopping until a clock line is raised. To avoid this problem, lines A and B have been added.

Line A is used to avoid a problem caused by overlap A in the timing chart, while line B solves the problem caused by overlap B.

This network is identical to ring No. 1 except for the additional freedom obtained in the clocking system.

## Ring Circuit No. 3

This network depicts a further modification that can be made in ring circuit No. 2 to enable it to operate from one clock line.

*Logic Network*

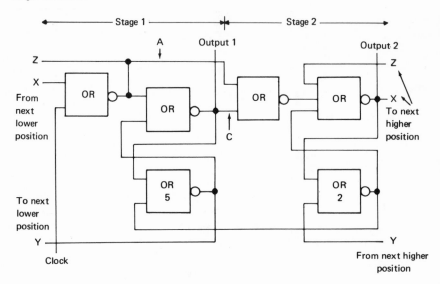

*Timing Chart (Four Outputs Shown)*

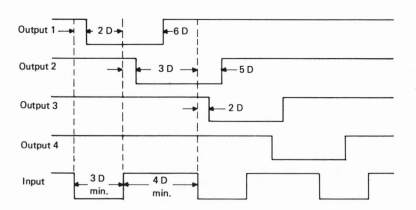

### Operational Notes

As explained in ring No. 2, a method has been found which permits the overlapping of the two clock pulses which drive gates 6 and 3. Now with this ability to tolerate a large variation in the clock timings, it is possible to look for a new source of pulses to replace clock line 2. This second clock signal can be obtained from the previous flip-flop. Lines A and C combine in gate 3 to permit the flow of a down condition from stage 1 to stage 2 when the clock line is raised. Line A is connected back to gate 1, forming a flip-flop from gates 6 and 1, to hold line A up for the negative period of the clock line.

### Restrictions

Network will not maintain down levels in adjacent stages.
Network can be used only when an even number of stages is tolerated.

### Additional Inputs

The raising of a normally down line connected to gates 2 and 5 will set all outputs to 1. When the clock line is up, the even number stages may be reset to 0 by raising a line connected to gate 1.

## Ring Circuit No. 4

All stages of this ring are identical in design and operation. This circuit enables the designer to form rings having an odd number of stages.

*Logic Network*

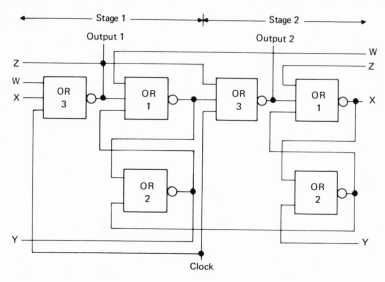

*Timing Chart (Three Stages Shown)*

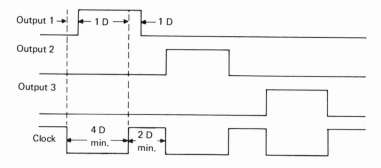

*Operational Notes*

The outputs of this ring are normally down, and an up level is propagate from stage to stage.

The ring advances on the fall of the clock line.

When the clock line is up, all outputs are down.

Normally down set and reset lines may be added to gates 2 and 1.

## Ring Circuit No. 5

This ring circuit is identical to the previous ring except it has been slightly modified to take advantage of a wired AND.

*Logic Network*

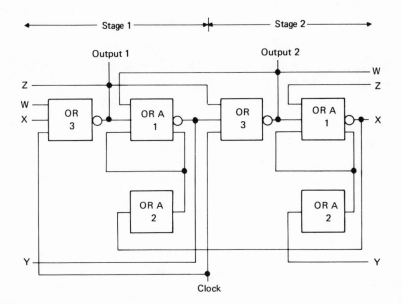

*Timing Chart*

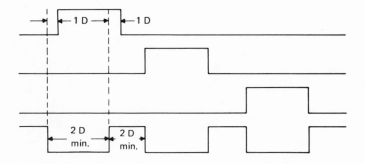

## Operational Notes

All outputs are normally down and an up level is propagated from stage to stage.

The ring advances on the fall of the clock line.

When the clock line is up, all outputs are down.

## Additional Inputs

Normally down set lines may be added to any gate 1. A normally up line added to gate 2 serves as a reset line.

# APPENDICES

# APPENDIX A

## Table of Powers of 2

| $2^n$ | n | $2^{-n}$ |
|---|---|---|
| 1 | 0 | 1.0 |
| 2 | 1 | 0.5 |
| 4 | 2 | 0.25 |
| 8 | 3 | 0.125 |
| 16 | 4 | 0.062 5 |
| 32 | 5 | 0.031 25 |
| 64 | 6 | 0.015 625 |
| 128 | 7 | 0.007 812 5 |
| 256 | 8 | 0.003 906 25 |
| 512 | 9 | 0.001 953 125 |
| 1 024 | 10 | 0.000 976 562 5 |
| 2 048 | 11 | 0.000 488 281 25 |
| 4 096 | 12 | 0.000 244 140 625 |
| 8 192 | 13 | 0.000 122 070 312 5 |
| 16 384 | 14 | 0.000 061 035 156 25 |
| 32 768 | 15 | 0.000 030 517 578 125 |
| 65 536 | 16 | 0.000 015 258 789 062 5 |
| 131 072 | 17 | 0.000 007 629 394 531 25 |
| 262 144 | 18 | 0.000 003 814 697 265 625 |
| 524 288 | 19 | 0.000 001 907 348 632 812 5 |
| 1 048 576 | 20 | 0.000 000 953 674 316 406 25 |
| 2 097 152 | 21 | 0.000 000 476 837 158 203 125 |
| 4 194 304 | 22 | 0.000 000 238 418 579 101 562 5 |
| 8 388 608 | 23 | 0.000 000 119 209 289 550 781 25 |
| 16 777 216 | 24 | 0.000 000 059 604 644 775 390 625 |
| 33 554 432 | 25 | 0.000 000 029 802 322 387 695 312 5 |
| 67 108 864 | 26 | 0.000 000 014 901 161 193 847 656 25 |
| 134 217 728 | 27 | 0.000 000 007 450 580 596 923 828 125 |
| 268 435 456 | 28 | 0.000 000 003 725 290 298 461 914 062 5 |
| 536 870 912 | 29 | 0.000 000 001 862 645 149 230 957 031 25 |
| 1 073 741 824 | 30 | 0.000 000 000 931 322 574 615 478 515 625 |
| 2 147 483 648 | 31 | 0.000 000 000 465 661 287 307 739 257 812 5 |
| 4 294 967 296 | 32 | 0.000 000 000 232 830 643 653 869 628 906 25 |
| 8 589 934 592 | 33 | 0.000 000 000 116 415 321 826 934 814 453 125 |
| 17 179 869 184 | 34 | 0.000 000 000 058 207 660 913 467 407 226 562 5 |

# APPENDIX B

## Decimal to Binary Conversion Table

| Decimal number | Binary number | Number of "1's" in binary number |
|:---:|:---:|:---:|
| 0 | 0 | 0 |
| 1 | 1 | 1 |
| 2 | 10 | 1 |
| 3 | 11 | 2 |
| 4 | 100 | 1 |
| 5 | 101 | 2 |
| 6 | 110 | 2 |
| 7 | 111 | 3 |
| 8 | 1000 | 1 |
| 9 | 1001 | 2 |
| 10 | 1010 | 2 |
| 11 | 1011 | 3 |
| 12 | 1100 | 2 |
| 13 | 1101 | 3 |
| 14 | 1110 | 3 |
| 15 | 1111 | 4 |
| 16 | 10000 | 1 |
| 17 | 10001 | 2 |
| 18 | 10010 | 2 |
| 19 | 10011 | 3 |
| 20 | 10100 | 2 |
| 21 | 10101 | 3 |
| 22 | 10110 | 3 |
| 23 | 10111 | 4 |
| 24 | 11000 | 2 |
| 25 | 11001 | 3 |
| 26 | 11010 | 3 |
| 27 | 11011 | 4 |
| 28 | 11100 | 3 |
| 29 | 11101 | 4 |
| 30 | 11110 | 4 |
| 31 | 11111 | 5 |
| 32 | 100000 | 1 |
| 33 | 100001 | 2 |
| 34 | 100010 | 2 |
| 35 | 100011 | 3 |
| 36 | 100100 | 2 |
| 37 | 100101 | 3 |
| 38 | 100110 | 3 |
| 39 | 100111 | 4 |
| 40 | 101000 | 2 |
| 41 | 101001 | 3 |

| Decimal number | Binary number | Number of "1's" in binary number |
|---|---|---|
| 42 | 101010 | 3 |
| 43 | 101011 | 4 |
| 44 | 101100 | 3 |
| 45 | 101101 | 4 |
| 46 | 101110 | 4 |
| 47 | 101111 | 5 |
| 48 | 110000 | 2 |
| 49 | 110001 | 3 |
| 50 | 110010 | 3 |
| 51 | 110011 | 4 |
| 52 | 110100 | 3 |
| 53 | 110101 | 4 |
| 54 | 110110 | 4 |
| 55 | 110111 | 5 |
| 56 | 111000 | 3 |
| 57 | 111001 | 4 |
| 58 | 111010 | 4 |
| 59 | 111011 | 5 |
| 60 | 111100 | 4 |
| 61 | 111101 | 5 |
| 62 | 111110 | 5 |
| 63 | 111111 | 6 |
| 64 | 1000000 | 1 |
| 65 | 1000001 | 2 |
| 66 | 1000010 | 2 |
| 67 | 1000011 | 3 |
| 68 | 1000100 | 2 |
| 69 | 1000101 | 3 |
| 70 | 1000110 | 3 |
| 71 | 1000111 | 4 |
| 72 | 1001000 | 2 |
| 73 | 1001001 | 3 |
| 74 | 1001010 | 3 |
| 75 | 1001011 | 4 |
| 76 | 1001100 | 3 |
| 77 | 1001101 | 4 |
| 78 | 1001110 | 4 |
| 79 | 1001111 | 5 |
| 80 | 1010000 | 2 |
| 81 | 1010001 | 3 |
| 82 | 1010010 | 3 |
| 83 | 1010011 | 4 |
| 84 | 1010100 | 3 |
| 85 | 1010101 | 4 |
| 86 | 1010110 | 4 |
| 87 | 1010111 | 5 |
| 88 | 1011000 | 3 |
| 89 | 1011001 | 4 |

| Decimal number | Binary number | Number of "1's" in binary number |
|:---:|:---:|:---:|
| 90 | 1011010 | 4 |
| 91 | 1011011 | 5 |
| 92 | 1011100 | 4 |
| 93 | 1011101 | 5 |
| 94 | 1011110 | 5 |
| 95 | 1011111 | 6 |
| 96 | 1100000 | 2 |
| 97 | 1100001 | 3 |
| 98 | 1100010 | 3 |
| 99 | 1100011 | 4 |
| 100 | 1100100 | 3 |

# APPENDIX C

## Decimal-Gray Code Table

| Dec | | | | | | | Dec | | | | | | | |
|---|---|---|---|---|---|---|---|---|---|---|---|---|---|---|
| 0 |   |   |   |   |   | 0 | 41 | 1 | 1 | 1 | 1 | 0 | 1 |
| 1 |   |   |   |   |   | 1 | 42 | 1 | 1 | 1 | 1 | 1 | 1 |
| 2 |   |   |   |   | 1 | 1 | 43 | 1 | 1 | 1 | 1 | 1 | 0 |
| 3 |   |   |   |   | 1 | 0 | 44 | 1 | 1 | 1 | 0 | 1 | 0 |
| 4 |   |   |   | 1 | 1 | 0 | 45 | 1 | 1 | 1 | 0 | 1 | 1 |
| 5 |   |   |   | 1 | 1 | 1 | 46 | 1 | 1 | 1 | 0 | 0 | 1 |
| 6 |   |   |   | 1 | 0 | 1 | 47 | 1 | 1 | 1 | 0 | 0 | 0 |
| 7 |   |   |   | 1 | 0 | 0 | 48 | 1 | 0 | 1 | 0 | 0 | 0 |
| 8 |   |   | 1 | 1 | 0 | 0 | 49 | 1 | 0 | 1 | 0 | 0 | 1 |
| 9 |   |   | 1 | 1 | 0 | 1 | 50 | 1 | 0 | 1 | 0 | 1 | 1 |
| 10 |   |   | 1 | 1 | 1 | 1 | 51 | 1 | 0 | 1 | 0 | 1 | 0 |
| 11 |   |   | 1 | 1 | 1 | 0 | 52 | 1 | 0 | 1 | 1 | 1 | 0 |
| 12 |   |   | 1 | 0 | 1 | 0 | 53 | 1 | 0 | 1 | 1 | 1 | 1 |
| 13 |   |   | 1 | 0 | 1 | 1 | 54 | 1 | 0 | 1 | 1 | 0 | 1 |
| 14 |   |   | 1 | 0 | 0 | 1 | 55 | 1 | 0 | 1 | 1 | 0 | 0 |
| 15 |   |   | 1 | 0 | 0 | 0 | 56 | 1 | 0 | 0 | 1 | 0 | 0 |
| 16 |   | 1 | 1 | 0 | 0 | 0 | 57 | 1 | 0 | 0 | 1 | 0 | 1 |
| 17 |   | 1 | 1 | 0 | 0 | 1 | 58 | 1 | 0 | 0 | 1 | 1 | 1 |
| 18 |   | 1 | 1 | 0 | 1 | 1 | 59 | 1 | 0 | 0 | 1 | 1 | 0 |
| 19 |   | 1 | 1 | 0 | 1 | 0 | 60 | 1 | 0 | 0 | 0 | 1 | 0 |
| 20 |   | 1 | 1 | 1 | 1 | 0 | 61 | 1 | 0 | 0 | 0 | 1 | 1 |
| 21 |   | 1 | 1 | 1 | 1 | 1 | 62 | 1 | 0 | 0 | 0 | 0 | 1 |
| 22 |   | 1 | 1 | 1 | 0 | 1 | 63 | 1 | 0 | 0 | 0 | 0 | 0 |
| 23 |   | 1 | 1 | 1 | 0 | 0 | 64 | 1 | 1 | 0 | 0 | 0 | 0 | 0 |
| 24 |   | 1 | 0 | 1 | 0 | 0 | 65 | 1 | 1 | 0 | 0 | 0 | 0 | 1 |
| 25 |   | 1 | 0 | 1 | 0 | 1 | 66 | 1 | 1 | 0 | 0 | 0 | 1 | 1 |
| 26 |   | 1 | 0 | 1 | 1 | 1 | 67 | 1 | 1 | 0 | 0 | 0 | 1 | 0 |
| 27 |   | 1 | 0 | 1 | 1 | 0 | 68 | 1 | 1 | 0 | 0 | 1 | 1 | 0 |
| 28 |   | 1 | 0 | 0 | 1 | 0 | 69 | 1 | 1 | 0 | 0 | 1 | 1 | 1 |
| 29 |   | 1 | 0 | 0 | 1 | 1 | 70 | 1 | 1 | 0 | 0 | 1 | 0 | 1 |
| 30 |   | 1 | 0 | 0 | 0 | 1 | 71 | 1 | 1 | 0 | 0 | 1 | 0 | 0 |
| 31 |   | 1 | 0 | 0 | 0 | 0 | 72 | 1 | 1 | 0 | 1 | 1 | 0 | 0 |
| 32 | 1 | 1 | 0 | 0 | 0 | 0 | 73 | 1 | 1 | 0 | 1 | 1 | 0 | 1 |
| 33 | 1 | 1 | 0 | 0 | 0 | 1 | 74 | 1 | 1 | 0 | 1 | 1 | 1 | 1 |
| 34 | 1 | 1 | 0 | 0 | 1 | 1 | 75 | 1 | 1 | 0 | 1 | 1 | 1 | 0 |
| 35 | 1 | 1 | 0 | 0 | 1 | 0 | 76 | 1 | 1 | 0 | 1 | 0 | 1 | 0 |
| 36 | 1 | 1 | 0 | 1 | 1 | 0 | 77 | 1 | 1 | 0 | 1 | 0 | 1 | 1 |
| 37 | 1 | 1 | 0 | 1 | 1 | 1 | 78 | 1 | 1 | 0 | 1 | 0 | 0 | 1 |
| 38 | 1 | 1 | 0 | 1 | 0 | 1 | 79 | 1 | 1 | 0 | 1 | 0 | 0 | 0 |
| 39 | 1 | 1 | 0 | 1 | 0 | 0 | 80 | 1 | 1 | 1 | 1 | 0 | 0 | 0 |
| 40 | 1 | 1 | 1 | 1 | 0 | 0 |   |   |   |   |   |   |   |

# APPENDIX D

**Decimal Table Grouped According to Number of "1's" in Binary Representation: Table from 1 to 127**

| One "1" | Three "1's" | Four "1's" | Five "1's" | Six "1's" |
|---|---|---|---|---|
| 1 | 7 | 15 | 31 | 63 |
| 2 | 11 | 23 | 47 | 95 |
| 4 | 13 | 27 | 55 | 111 |
| 8 | 14 | 29 | 59 | 119 |
| 16 | 19 | 30 | 61 | 123 |
| 32 | 21 | 39 | 62 | 125 |
| | 22 | 43 | 79 | 126 |
| | 25 | 45 | 87 | |
| **Two "1's"** | 26 | 46 | 91 | |
| | 28 | 51 | 93 | **Seven "1's"** |
| 3 | 35 | 53 | 94 | |
| 5 | 37 | 54 | 103 | 127 |
| 6 | 38 | 57 | 107 | |
| 9 | 41 | 58 | 109 | |
| 10 | 42 | 60 | 110 | |
| 12 | 44 | 71 | 115 | |
| 17 | 49 | 75 | 117 | |
| 18 | 50 | 77 | 118 | |
| 20 | 52 | 78 | 121 | |
| 24 | 56 | 83 | 122 | |
| 33 | 67 | 85 | 124 | |
| 34 | 69 | 86 | | |
| 36 | 70 | 89 | | |
| 40 | 73 | 90 | | |
| 48 | 74 | 92 | | |
| 65 | 76 | 99 | | |
| 66 | 81 | 101 | | |
| 68 | 82 | 102 | | |
| 72 | 84 | 105 | | |
| 80 | 88 | 106 | | |
| 96 | 97 | 108 | | |
| | 98 | 113 | | |
| | 100 | 114 | | |
| | 104 | 116 | | |
| | 112 | 120 | | |

# APPENDIX E

## Theorems in Boolean Algebra

1.    $X \cdot 0$     $=$     $0$
2.    $X \cdot 1$     $=$     $X$
3.    $X \cdot X$     $=$     $X$
4.    $X \cdot \overline{X}$     $=$     $0$
5.    $X + 0$     $=$     $X$
6.    $X + 1$     $=$     $1$
7.    $X + X$     $=$     $X$
8.    $X + \overline{X}$     $=$     $1$
9.    $\overline{\overline{X}}$     $=$     $X$
10.   $X \cdot Y$     $=$     $Y \cdot X$
11.   $X + Y$     $=$     $Y + X$
12.   $X \cdot (Y \cdot Z) = (X \cdot Y) \cdot Z = X \cdot Y \cdot Z$
13.   $X + (Y + Z) = (X + Y) + Z = X + Y + Z$
14.   $X \cdot Y + X \cdot Z = X \cdot (Y + Z)$
15.   $(X + Y) \cdot (X + Z) = X + Y \cdot Z$
16.   $X \cdot Y + X \cdot \overline{Y} = X$
17.   $X + X \cdot Y = X + Y$
18.   $X + \overline{X} \cdot Y = X + Y$
19.   $X \cdot Y + \overline{X} \cdot Z + Y \cdot Z = X \cdot Y + \overline{X} \cdot Z$
20.   $(X + Y) \cdot (X + Z) \cdot (Y + Z) = (X + Y) \cdot (\overline{X} + Z)$
21.   $\overline{X \cdot Y} = \overline{X} + \overline{Y}$
22.   $\overline{X + Y} = \overline{X} \cdot \overline{Y}$

### *Sum Modulo Two*

23.   $X \mathbin{\veebar} 0$     $=$     $X$
24.   $X \mathbin{\veebar} 1$     $=$     $\overline{X}$
25.   $X \mathbin{\veebar} X$     $=$     $0$
26.   $X \mathbin{\veebar} \overline{X}$     $=$     $1$
27.   $X \mathbin{\veebar} Y$     $=$     $Y \mathbin{\veebar} X$
28.   $X \cdot Y \mathbin{\veebar} X \cdot Z = X \cdot (Y \mathbin{\veebar} Z)$
29.   $X \mathbin{\veebar} Y$     $=$     $X \cdot \overline{Y} + \overline{X} \cdot Y$
30.   $X \mathbin{\veebar} \overline{Y}$     $=$     $X \cdot Y + \overline{X} \cdot \overline{Y}$
31.   $\overline{X \mathbin{\veebar} Y}$     $=$     $X \mathbin{\veebar} \overline{Y} = \overline{X} \mathbin{\veebar} Y$

## APPENDIX F

**Error Correcting Code (Hamming)**

*Single Error Correcting Code*

The general equation for the number of check bits "k" required to encode "n" information bits is:

$$2k = n + k + 1$$

Examples:

| Information Bits | Check Bits Required |
| --- | --- |
| 1 | 2 |
| 2 through 4 | 3 |
| 5 through 11 | 4 |
| 12 through 26 | 5 |
| 27 through 57 | 6 |

A simple method for determining the number of check bits required and the bit positions covered by each check bit is shown below.

1. Start a counting sequence of numbers starting at one as shown:

   1, 2, 3, 4, 5, 6, 7, 8, 9, 10, 11, 12, 13, 14, 15, etc.

2. These numbers represent positions in the Hamming code and the power of 2 positions (1, 2, 4, 8, etc.) are reserved for check bits. All other positions are reserved for data

   ```
   1  2  3  4  5  6  7  8  9 10 11 12 13 14 15 etc.
   C  C     C           C
            D     D  D  D     D  D  D  D  D  D  D etc.
   ```

3. Check bit 1 is used to obtain an even check on all positions that have a one in the binary representation of their position number, as shown by the X's below:

```
1  2  3  4  5  6  7  8  9 10 11 12 13 14 15 etc.
C  C     C           C
      D     D  D  D     D  D  D  D  D. D  D
X     X  X  X     X  X  X     X
```

4. Check bit 2 is used to obtain an even check on all positions that have a two in the binary representation of their position number as shown by the Y's below:

```
1  2  3  4  5  6  7  8  9 10 11 12 13 14 15 etc.
X     X  X     X  X     X  X     X  X
   Y  Y        Y  Y        Y  Y
```

5. All other check bits follow this same procedure.
   Thus:
   Check bit 4 covers positions 4, 5, 6, 7, 12, 13, 14, 15, etc.
   Check bit 8 covers positions 8, 9, 10, 11, 12, 13, 14, 15, etc.
   To correct a word:

   (a) Compare check bits which arrived with word against check bits generated from the data bits following the above rules.

   (b) If these two sets of check bits are identical the word is correct.

   (c) If these two sets of check bits disagree then record the position numbers of the disagreeing bits. The sum of these position numbers indicates which bit position is wrong. Thus if check bit positions 1 and 4 disagree, position 5 is in error and should be inverted.

# APPENDIX G

## Circuit Types

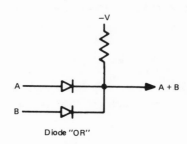

Diode "OR"

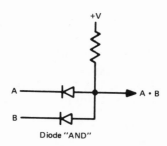

Diode "AND"

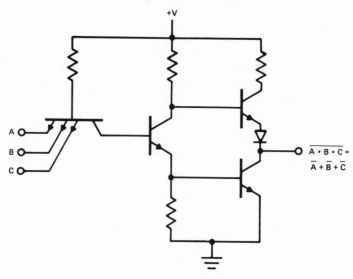

TTL or T$^2$L          Transistor-Transistor Logic

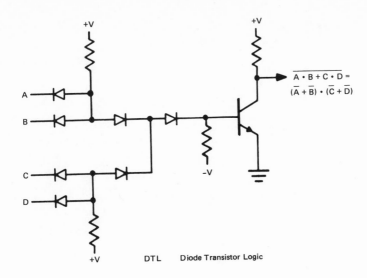

DTL    Diode Transistor Logic

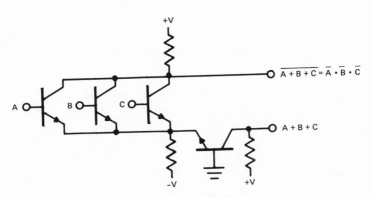

ECL    Emitter Coupled Logic

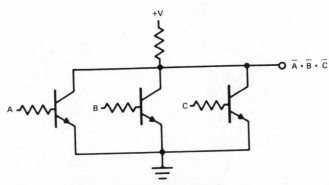

RTL = Resistor Transistor Logic

# INDEX